Encounter with the Earth
Materials and Processes

Encounter with the Earth
Materials and Processes

Léo F. Laporte
University of California, Santa Cruz

Canfield Press ΦP San Francisco
A Department of Harper & Row, Publishers, Inc.
New York   London

Sponsoring editor: Howard Boyer
Production editor: Pat Brewer
Artists: Larry Jansen, Pam Vesterby
Designer: Barbara Robinson
Copyeditor: Vickie Golden
Photo Researcher: Kay Y. James
Cover photograph: U.S. Geological Survey

Library of Congress Cataloging in Publication Data

Laporte, Léo F.
  Encounter with the earth.

  Includes bibliographies and indexes.
  CONTENTS: v. 1. Materials and processes.—v. 2. Resources.—v. 3. Wastes and
hazards.
  1. Earth sciences. 2. Natural resources. 3. Pollution. I. Title.
QE28.L24   1975b   550   75-23034
ISBN 0-06-384782-5 (v. 1.)

This book was previously published as Part One of
Encounter with the Earth by Léo F. Laporte.

Encounter with the Earth: Materials and Processes

75 76 77 10 9 8 7 6 5 4 3 2 1

Credits begin on p. 175.

Contents

Contents

Preface

This book provides a core summary of basic geologic processes. In the first two chapters we discuss the macroscopic and microscopic materials of the earth. In the next three chapters we review the internal, surface, and biological processes that keep the earth in a constant state of change. In the last chapter we explore temporal aspects of the earth: its long history that has brought us to where we are today, and how processes operate over different intervals of time. The emphasis is on giving a coherent, concise, and systematic understanding of what our spaceship is and how it works.

Several unique elements have been incorporated in the book. Some of the social, economic, and political issues that are inevitably entangled with environmental concerns are introduced, and in some instances discussed at length. Each chapter includes a Viewpoint written by an authority in the field to expand on some element treated in the chapter or, in a few cases, to introduce another aspect or opinion. These Viewpoints were written to give the reader a sense of the ongoing search for strategies and tactics for living on a finite earth and to avoid cut-and-dried statements on particular issues or problems. Marginal notes throughout each chapter refer forward and backward to other sections of the book that relate to the discussion. Some marginal notes suggest topics for more open-ended discussions as well as projects that might be pursued. In addition, a summary, glossary, and list of suggested readings at the end of each chapter should be of use to the student.

No textbook is written in a vacuum. Of necessity an introductory book that covers a wide range of subjects requires the generous help of many people. In particular, I wish to acknowledge the assistance of the many people who contributed opinions, ideas, or information as the book progressed. I am grateful to the following persons who commented on the initial organization and aims of the book: Robert E. Boyer, University of Texas, Austin; Edward A. Hay, De Anza College; Bernard W. Pipkin, University of Southern California; Arthur N. Strahler, University of California, Santa Barbara; and J. T. Wilson, University of Toronto.

The full text was reviewed by the following persons, none of whom, of course, is reponsible for any misstatements or inaccuracies: John G. Dennis, California State University, Long Beach; Richard R. Doell, U.S. Geological Survey, Menlo Park; M. Grant Gross, Chesapeake Bay Institute; N. Timothy Hall, Foothill College; Charles B. Hunt, Johns Hopkins University; John H. Moss, Franklin and Marshall College; Edward O'Donnell, University of South Florida; Kazimierz M. Pohopien, Mt. San Antonio College; Robert L. Rose, California State University, San Jose; and Henry I. Snider, Eastern Connecticut State College. I appreciate the effort of Harold C. Urey, University of California, San Diego, who reviewed Chapter 1. Special thanks are given to the six persons who found time to submit interesting and insightful Viewpoints and thereby broaden the appeal and range of the text, and in particular, to my colleague, Gerald Bowden, for his trenchant wit in the accompanying cartoons.

Anyone who has written a book knows how crucially important the knowledge, skills, and patience of editors are. So I thank Pat Brewer, production editor, for her talent in putting it all together with imagination and taste. I cannot express fully my appreciation to Howard Boyer, project editor, who saw this book through from vague notions to completed text.

Santa Cruz, California
June 1975

Léo F. Laporte

Encounter with the Earth
Materials and Processes

Materials and Processes

Molten lava lake, Hawaiian Islands.

The View from Space

The view of earth from space—a unique, all-encompassing perspective of our human habitat—complements and enhances our everyday, earthbound view. As we sit here with our ant's eye view of the globe, our natural environment seems to extend in all directions, up, down, and sideways. The air we breathe and the water we drink lie all around us, as do the rocks and soil that provide us with minerals, fuels, food, and fiber. What we lack here is available over the next hill.

But our recent views of earth from space show the earth as a limited, lonely wanderer:

- Earth is the only planet inhabited by intelligent life within the solar system.
- All life occurs within the biosphere, a thin film at the earth's surface.
- This biosphere is limited in its abundance and distribution by three other similarly thin shells of atmospheric gas, surface water, and solid crust.
- The physical and chemical properties of these thin spherical shells are not typical of the earth as a whole, or of the solar system, or of our galaxy.
- The special character of these shells, upon which my life, your life, and all life depend, results from the original formation of the earth and its subsequent development.

Thus, what appeared limitless and without bounds has become limited and bounded. Paradoxically, the view from space restricts our spacious view.

This rocky orb with its life-sustaining fluid and solid shells is the ultimate determinant of the events unfolding upon it—past, present, and future. We ride with it as it hurtles through space in tandem with the sun, its sister planets, and the rest of the galactic swarm. In this initial chapter we want to hold our earth at arm's length, as it were, examining its structure

Once a photograph of the Earth, taken from outside, is available, once the sheer isolation of the Earth becomes plain, a new idea as powerful as any in history will be let loose.

Fred Hoyle, 1948

Cloud's Viewpoint at the end of the chapter takes such a global perspective of our spaceship.

and composition, as well as discovering how it came to be.

We begin with a summary of the earth's formation; the development of its internal, solid shells of core, mantle, and crust; and the accompanying accumulation of the fluid, outer shells of airy atmosphere and watery hydrosphere. Finally, we arrive at the origin of the biosphere: that marvelously tenuous yet tenacious realm of life that exists at the interface of solid and fluid earth.

1-1 How Primitive Earth Formed

As solid and unchanging as the earth may seem, it was not always so. There was actually a time, billions of years ago, when the earth simply did not exist; nor did the sun, moon, or other planets. Instead, the solar system—and probably other stars within our Milky Way galaxy—was dispersed as a huge, diffuse mass of gas and tiny solid particles. Only over the course of eons did this dust cloud slowly condense to become our sun with its nine planets. Even then, however, the earth was not as it is now. Another long interval had to elapse before this planet developed into the layered body we see today: an inner, iron-nickel core surrounded by a mantle of rock with a thin, surface crust. The processes that led to the layering also released the gases and liquids that became the early atmosphere and hydrosphere. And it was at this point, after a few billion years of development, that the stage was set for the origin and evolution of life that composes the biosphere.

The center of the Milky Way galaxy as viewed from our solar system at its outer edge. The Milky Way is composed of galactic dust, gas, and stars, one of which is our sun.

Some chemical terms

Before proceeding further, we need to introduce and briefly define a few chemical terms. All the substances of the universe are composed of one or more *elements*; for example, water is composed of the two elements hydrogen and oxygen. Each of the 106 elements has a one- or two-letter symbol; to continue our example, hydrogen's symbol is H and oxygen's is O. An *atom* is the basic unit of an element—if we cut an atom apart, we no longer have the element. Within all atoms there are at least two different kinds of particles: protons and electrons; most atoms also have neutrons. A *proton* is a positively charged mass found in the central portion, or *nucleus*, of the element. An *electron* is a negatively charged particle whose mass is about 1/1800 that of a proton; it travels in an "orbit" around the nucleus. A *neutron* has no charge and has a mass about equal to that of a proton; with the proton it forms the nucleus. The number of protons within the nucleus determines an element's *atomic number*, while the number of protons and neutrons (if any) define its *atomic mass*. Oxygen, for example, has an atomic number of 8 and an atomic mass of 16 because it has 8 protons and 8 neutrons within its nucleus. Hydrogen, on the other hand, has an atomic number of 1 and a mass of 1 because it has just a single proton in its nucleus. The first twenty elements, with their symbols and atomic numbers, are listed in Table 1–1.

The electrons surround the nucleus in rapidly moving orbits. When the number of electrons equals the number of protons, an atom is said to be electrically balanced or neutral. However, electrons may be added to or removed from an atom to make it electrically negative or positive. Atoms that have gained or lost one or more electrons are called *ions*. We denote ions with a superscript next to the element's symbol—a plus sign ($+$) if the atom has lost electrons, a minus sign ($-$) if it has gained electrons. A number is used if the atom has gained or lost more than one electron. For example, if the oxygen atom gains two electrons, the oxygen ion is O^{2-}; if the sodium atom loses one electron, its ion is Na^+.

When two or more elements chemically combine to form a *compound*, they then share electrons. A *molecule* is the smallest unit of a chemical compound that still has the unique properties of that compound. A molecule of water (H_2O) contains one oxygen ion (O^{2-}) and two hydrogen ions (H^+); the two hydrogen ions in the molecule of water are indicated by the subscript 2 in H_2O. Salt (NaCl) has one sodium ion (Na^+) and one chlorine ion (Cl^-). Note that a water molecule and a salt molecule are electrically balanced.

Elements can be classified by their properties and behaviors. One classification that we will be using is metals, nonmetals, and metalloids. A *metal* is opaque, has a characteristic luster, conducts electricity and heat, and can be fused. A *nonmetal* lacks some or all of these properties. A

Table 1–1   First Twenty Elements

Atomic Number	Element's Name	Chemical Symbol
1	hydrogen	H
2	helium	He
3	lithium	Li
4	beryllium	Be
5	boron	B
6	carbon	C
7	nitrogen	N
8	oxygen	O
9	fluorine	F
10	neon	Ne
11	sodium	Na
12	magnesium	Mg
13	aluminum	Al
14	silicon	Si
15	phosphorus	P
16	sulfur	S
17	chlorine	Cl
18	argon	Ar
19	potassium	K
20	calcium	Ca

Recall that an electron is negatively charged.

metalloid has some metallic and some nonmetallic characteristics. Aluminum is a metal, carbon is a nonmetal, and silicon is a metalloid.

A compound composed of metallic (or nonmetallic) ions and oxygen ions is called an *oxide*—Fe_2O_3 is an iron oxide. The *silicates* are a special category of compounds composed of two or more oxides, one of which always includes silicon dioxide (SiO_2)—Al_2SiO_5 is aluminum silicate. As we will see shortly, most of the earth's crust is composed of oxides, especially the silicates.

Planetary formation

Although important details on the origin of the solar system are still in doubt, many astronomers and earth scientists agree that the sun and its planets condensed from a *nebula*, a large rotating mass of gaseous and solid material. Starting almost 5 billion years ago, this nebula slowly condensed and contracted, gradually forming a flattened disk whose center became the sun. The sun's composition, mainly hydrogen and helium (He), with lesser amounts of oxygen, carbon (C), and the other elements, is proportionately similar to the abundance of these elements in our Milky Way galaxy. The surrounding disk, containing about one-third the mass of the original nebula and with a composition like the sun's, consisted of particles orbiting the sun.

Within about 100 million years, the material in the disk contracted into increasingly larger, discrete, aggregated bodies that eventually became the planets, asteroids, comets, and moons. The mechanism responsible for this aggregation is not clear, although one theory suggests that the planets formed within eddies in the rotating disk. This hypothesis is shown in Figure 1–1.

The four inner planets closest to the sun—Mercury, Venus, Earth, and Mars—were apparently too small in mass to retain their more volatile substances, like hydrogen and helium. As these planets were heated by the sun and later by radioactive decay of some of their elements, the volatile substances escaped the planets' gravitational pull. Hence, the inner planets have a chemical composition different from that of the original nebula, as now found in the sun. These planets are rich in nonvolatile, high-density materials in the form of oxides of iron, silicon, magnesium, and other elements. The extreme chemical activity of oxygen kept this element bound to the nonvolatile metals rather than being dispersed as an elemental gas, as in the case of hydrogen and helium, or as water.

These are known as the terrestrial planets.

These are called the Jovian planets.

The more massive outer planets—Jupiter, Saturn, Uranus, and Neptune—kept some of their volatile helium and hydrogen because of their greater gravitational forces. Having retained the average composition of the original nebula, these planets are more like the sun, with large amounts of the lower density volatiles and smaller amounts of carbon, nitrogen (N), and oxygen atoms in the form of methane (CH_4), ammonia (NH_3), and water

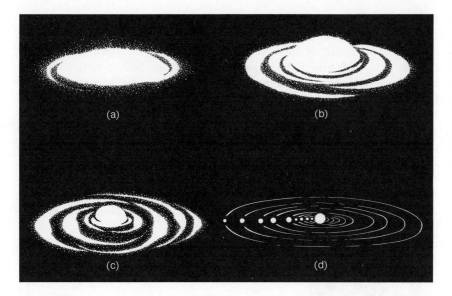

Figure 1-1

Nebular hypothesis of the solar system's origin. A huge cloud of gas and dust (a) slowly contracts and condenses into a flattened disk with the sun at the center (b). Planets begin to accrete or accumulate within the disk (c), so that eventually the nebula evolves into the sun with nine planets revolving around it (d).

(H_2O). Because of the considerable gravitational mass and low temperatures of the outer planets, all these substances are present either as solids or extremely dense gases that behave like solids. Pluto, the outermost planet, does not resemble its four outer planetary companions and remains a puzzle to astronomers. One conjecture is that Pluto is a lost satellite of Neptune.

Differentiation through heating

Within about 100 million years, the earth condensed and aggregated into a cold mass of rock rich in metallic oxides with no significant atmosphere or hydrosphere. After this initial period of relatively cool accretion about 5 billion years ago, the earth slowly began to grow warmer. The heat came from several sources. As additional planetary particles, some several hundred kilometers in diameter, collided with the accreting early-earth, the impact generated heat. In addition, the early-earth's gravitational forces compacted and consolidated the growing planetary mass, generating still more heat. The decay of radioactive elements like uranium, thorium, and potassium was a third source of heat.

Similar heat is generated when you clap your hands vigorously.

As the temperature of the early-earth increased, the melting point of many of its constituents was reached. Heating and melting within the early-earth marked a new and important stage in the earth's development: its concentric differentiation into a layered body containing core, mantle, and crust.

The individual particles that accumulated to form the early-earth each had the same general composition. They were rich in iron (35 percent by weight); oxygen (30 percent); silicon (15 percent); magnesium (13 percent); and about 1–2 percent each of nickel, sulfur, calcium, and aluminum—accounting for about 99.5 percent in all. Their distribution is described in Figure 1–2. At low temperatures, these materials existed either as uncombined elements or as metallic oxides and were evenly distributed. As the early-earth heated, they melted and formed new chemical compounds which were redistributed. Depending upon their chemical potential to react with each other, these substances formed a series of concentric shells decreasing in density from the interior outward. Thus, in about one-half billion years, the earth's core, mantle, and crust were created; Figure 1–3 illustrates the resulting layers.

A peach is a good analogy for this layering: skin, pulp, and pit within a pit. But are the proportions the same?

Figure 1–2

Major elements in the universe, sun, and earth in weight percent. The composition of the nebula that evolved into the solar system is assumed to have been similar to interstellar gas and dust observable in the universe today.

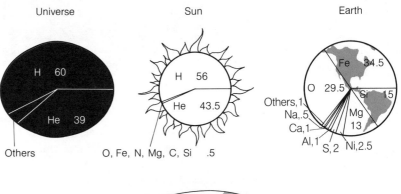

Figure 1–3

Cutaway view of the earth's interior. Core, mantle, and crust form concentric shells whose chemical composition became differentiated after initial cold accretion of the early-earth.

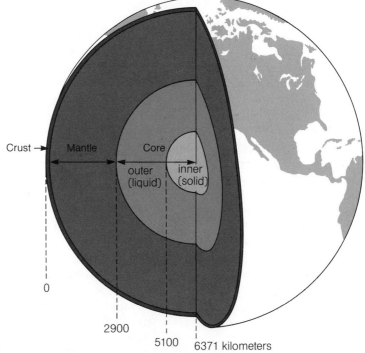

Core.   We cannot directly observe the composition and physical state of the earth's core. However, indirect evidence based on knowledge of the earth's overall density, measurements of earthquake waves that have passed through the earth, and results of laboratory experiments, indicates that the core is a mixture of iron and nickel about 3500 km in radius. The inner half is solid and the outer half is liquid. Rotation of the earth slowly moves the liquid portion, which in turn is believed to generate the magnetic field that surrounds the earth—in the same way a rotating dynamo creates electric currents which produce a magnetic field.

One kilometer is equal to 1000 meters, or about 5/8 mile; one mile is thus equivalent to 1.6 kilometers.

Mantle.   The earth's mantle, almost 2900 km thick, surrounds the core and consists of silicon oxides rich in iron and magnesium (89 percent) or sodium and aluminum (11 percent). The mantle has two portions, upper and lower, with the lower nearest the core. The upper mantle is near its melting point; although it is solid, it tends to react plastically, like "molasses in January." The lower mantle is more elastic—that is, like a hard rubber ball. Plastic behavior in the mantle occurs in response to stresses applied over long periods of time; elastic behavior results from short-term stresses. (Such combined elastic and plastic behavior is seen in "silly putty" that bounces elastically like a ball, but if left overnight settles slowly into a puddle.)

Crust.   The crust is the thin, outer shell of the earth. It is thinnest under the oceans (about 8 km) and thickest on the continents (about 40 km). While containing many of the substances found within the mantle, the crust is also enriched in potassium-aluminum silicates. In the next chapter we will discuss the great variety of rocks and minerals within the crust, and in subsequent chapters the occurrence and uneven distribution of mineral resources, oil and gas, and soil types. By contrast, the core and mantle are each much more homogeneous in chemical composition.

It's from the crust, of course, that we derive our metallic and nonmetallic resources.

Within the oceans the crust is thin, rich in iron-magnesium silicates, and covers about 53 percent of the earth's surface. Toward the margins of the continents the crust thickens and contains rocks with more sodium-aluminum, calcium-aluminum, and potassium-aluminum silicates. Within the continents the crust is thick (five times that under the oceans) and greatly enriched in various aluminum silicates. The continental crust covers about 29 percent of the earth (Figures 1–4 and 1–5).

The varying amounts of elements found in the earth's crust are indicated in Figure 1–4. Compare the weight percent data with Figure 1–2 and note how the crust is enriched in oxygen, silicon, aluminum, calcium, sodium, and potassium, and depleted in iron and magnesium, when related to the earth as a whole. The enriched elements are combined in a variety of silicate minerals such as quartz, feldspars, clays, amphiboles, and pyroxenes, as discussed in the next chapter.

Figure 1–4

Major elements in the earth's crust. Data shown by weight and volume. Notice, for example, that because of the relatively large volume of the oxygen atom almost 95 percent of the earth's crust by volume is composed of this one element! (Thickness of crust shown here is not to scale.)

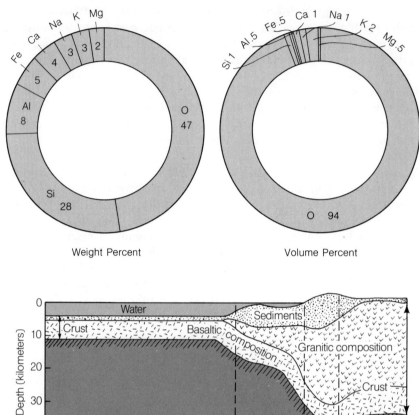

Weight Percent   Volume Percent

Figure 1–5

Cross section of the earth's crust showing dominant rock types, thickness, and area (in millions of square kilometers and percent).

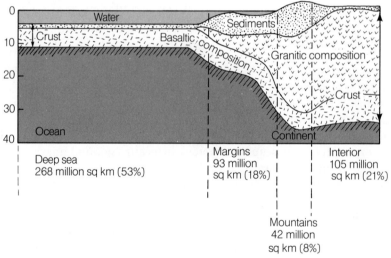

Deep sea
268 million sq km (53%)

Margins
93 million
sq km (18%)

Interior
105 million
sq km (21%)

Mountains
42 million
sq km (8%)

Thickness and composition of the crust and mantle are compared in Figure 1–5; the mantle has an average density or weight per unit volume of more than 3, whereas the crust has a density less than 3.0. Thus, where the crust is thicker, as in mountain regions, it "floats" higher—this is comparable to the idea of the larger the iceberg, the more its tip shows above the sea and the deeper it rides the water, as explained in Section 3–5. Chemical differentiation concentrated the lighter, more easily melted potassium-aluminum and sodium-aluminum silicates in the crust, while the heavier, less fusible, iron-magnesium silicates sank and accumulated within the mantle; see Figure 1–6.

1-2 How the Atmosphere and Hydrosphere Formed

During the period of planetary accretion, the earth virtually lacked the fluid shells of air and water that we call the atmosphere and hydrosphere. It wasn't until differentiation of the solid earth released hot gases from the planet's interior that the primitive atmosphere and hydrosphere started to accumulate in abundance. Once these fluid shells developed, the earth entered another stage in its long history. The presence of air and water initiated weathering and erosion of its surface rocks, as well as transportation and deposition of sediments; these processes released additional chemical elements to the accumulating fluid shells of air and water.

Escape from the interior

During the hot, chemical differentiation of the earth, radioactive elements—particularly uranium, thorium, and potassium—concentrated in the upper mantle and crust. The radioactive elements are unstable; that is, they decay spontaneously into stable elements, generating heat in the decay process. The time required for a radioactive substance to decay to one-half its original amount is called a *half-life*, and it varies for these three elements. But all the half-lives are sufficiently long (hundreds of millions to several billions of years) to create considerable heat throughout the earth's history. Figure 1–6 illustrates the rate of heat production from these radioactive elements for the last 5 billion years. One-half of the total radioactive heat was generated in the first billion and a half years; more than three-quarters was produced in the first half of the earth's existence.

The release of radioactive heat warmed various portions of the mantle and crust, causing them to melt and rise upward toward the surface. Mantle and crustal rocks brought into regions of lower pressure or higher temperature by this vertical movement formed *magmas*; that is, molten rock that penetrated the cooler solid rocks near the earth's surface, often breaking through as lava-producing volcanoes. These magmas also released gases composed mainly of water vapor; smaller amounts of carbon monoxide (CO) and dioxide (CO_2); chlorine (Cl); and traces of nitrogen (N), sulfur (S), and hydrogen (H). Some of the gases accumulated and formed the primitive atmosphere or condensed as liquids and created the primitive hydrosphere.

We can visualize this release of gases during the earth's hot chemical differentiation by a homely analogy. When an apple pie is put in the oven, it goes in cold. As it cooks, the liquids within the pie—mostly water—boil off. These gases—mostly steam—push up on the pie crust, making small, low domes on the pie surface or breaking through the crust. To prevent unattractive tears in the finished pie, any competent pastry cook will cut the pie crust before putting the pie in the oven to allow

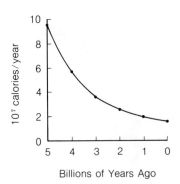

Figure 1–6

Rate of production of radioactive heat by uranium, potassium, and thorium during the almost 5-billion-year history of the earth. Notice that most of the radioactive heating occurred in the first third of earth history; it supplied much of the heat required for earth differentiation following initial planetary accretion.

This heat is responsible for geothermal energy.

11

Mt. Sakwiajima, Japan, venting gases from the earth's interior. Similar volcanic activity over the eons helped form the earth's atmosphere and hydrosphere.

these gases to escape during cooking. Since no cosmic pastry cook cut slits in the earth as it heated, volcanoes broke through the crust to release gases. The escaping gases had been chemically bound within the particles that accreted to form the early-earth; during hot differentiation and volcanism, they generated an early atmosphere and hydrosphere of carbon monoxide and dioxide, water vapor, hydrogen, nitrogen, and minor amounts of methane and ammonia. These chemical constituents provided the building blocks for life's origin and early evolution.

With the later appearance of plants, some 3 billion years ago, photosynthesis began to contribute significant amounts of oxygen to the atmosphere. As this oxygen accumulated, it combined eventually with the existing methane, ammonia, and carbon monoxide to produce carbon dioxide, nitrogen, and water. Thus evolved our present-day atmosphere rich in nitrogen and oxygen, with much smaller amounts of carbon dioxide. Table 1–2 provides a balance sheet of the additions and losses to the atmosphere during earth's history. Over a period of several billion years, a state of balance, or *dynamic equilibrium*, has been achieved with additions and losses roughly balancing out. More recently, however, human combustion of fossil fuels (coal, oil, and gas) has begun to add significant amounts of CO_2 to the atmosphere.

Why the sea is salty

The hydrosphere, unlike the atmosphere, is a discontinuous shell around the earth and includes the oceans, lakes, rivers, snow, ice, and underground water. It is also more variable in composition—especially as regards dissolved salts—ranging from pure rainwater to marine brines. Approximately 98 percent of the hydrosphere's volume is seawater, and so we discuss the

Table 1–2 Atmospheric Balance Sheet

Additions to the Atmosphere	Losses from the Atmosphere
Volcanoes	Oxidation of various chemical compounds removes oxygen
Oxygen from photochemical splitting of water molecules in the upper atmosphere	Fossil fuels and buried organic matter remove carbon dioxide
Oxygen from photosynthesis by plants	Formation of $CaCO_3$ and $MgCO_3$ removes carbon dioxide
Helium from radioactive decay of uranium and thorium	Nitrogen lost by formation of nitrogen oxides in atmosphere and nitrogen-using bacteria
Argon from radioactive decay of potassium	Helium and hydrogen escaping from the earth's gravitational field
Carbon dioxide from animals and plants	

origin and development of the oceans here. Later chapters treat other parts of the hydrosphere, particularly surface and underground water on the lands.

See Chapter 4 for a discussion of fresh water.

We assume that the early seas were less salty than they are today, because for billions of years running streams and rivers brought huge quantities of dissolved substances to the oceans. The primitive seas were also probably more acidic because of the greater amount of carbon dioxide in the early atmosphere. (Carbon dioxide in water forms a weak acid, H_2CO_3, carbonic acid.) But as sodium, magnesium, calcium, and aluminum eroded from the rocks on land, they gradually neutralized the acidic seas, making them slightly alkaline today.

One early attempt to date the earth's age involved dividing the total salts in the oceans by the amount added each year. As it turned out the age was too low, almost 50-fold. Can you guess why?

On a worldwide basis, seawater has about 35 parts of dissolved salts for each thousand parts of seawater. The major chemical substances in seawater that account for its salinity are chlorine, sodium, sulfate, magnesium, calcium, and potassium. Together they make up more than 99 percent of all the dissolved salts, as indicated in Table 1–3. Bicarbonate,

Table 1–3 Major Dissolved Substances in Seawater

Ion	Parts per Thousand	Percent
chloride (Cl^-)	19	55
sodium (Na^+)	11	31
sulfate (SO_4^{2-})	3	8
magnesium (Mg^{2+})	1	3
calcium (Ca^{2+})	½	1
potassium (K^+)	½	1
Total	35	99

bromine, boric acid, strontium, and fluorine contribute another 0.6 percent; the remaining chemical elements in seawater appear in trace amounts and account for less than one-tenth of one percent of the total.

In shallow areas of the sea where river runoff from the land is great, as in the Baltic Sea, salinities may be quite low—down to a few parts per thousand. In places where there are high evaporation rates, as in the Red Sea, salinities may be higher—about 40 parts per thousand. However, within most of the ocean, salinities vary only a part or two from the average 35 parts per thousand.

Various theories have been proposed regarding the rate of oceanic accumulation and salt concentration. Most agree, however, that the oceans have had approximately the same volume and salinity for about the last billion years. During most of that time a dynamic equilibrium has existed, so that the amount of dissolved material flowing annually to the oceans is about equal to the amount buried within their sediments. Also, the water falling in precipitation is balanced by the amounts which evaporate from their surface. Volcanoes add "new" water from the mantle and crust to the oceans, but this probably is compensated for by the amounts lost in various water-bearing minerals buried within marine sediments.

1-3 How the Biosphere Formed

Although the biosphere was the last to appear in earth history, the evolution of that thinnest and most tenuous of all earth shells is intimately related to the formation of the atmosphere and hydrosphere. Despite the great diversity of animal and plant life, organisms are remarkably similar in their chemical composition. Table 1–4 details how they all contain compounds rich in hydrogen, carbon, nitrogen, and oxygen—the very same elements that were abundant in the primitive atmosphere and hydrosphere. These four elements account for more than 90 percent of living matter and are combined into a variety of large molecules that are the building blocks for all organisms. The remaining few percent of living matter is composed of a number of elements including magnesium, calcium, iron, potassium, phosphorus, and silicon.

These few elements combine to form an incredible variety of vital compounds including amino acids, proteins, starches, and fats.

Photosynthesis

Very early in the earth's history, energy supplied from the sun's ultraviolet radiation and lightning created several complex organic molecules in the sea. These molecules became the precursor substances for later development of the simplest kinds of life, which resembled single-celled algae and bacteria. By 3 billion years ago, the important biological process of photosynthesis had evolved among some of these earliest forms of life.

14

Table 1–4 Chemical Composition of Organic Matter

Atom	Molecule	Vital Compound	Weight Percent Plant	Animal
hydrogen	acid ⟶	lipids, fats	1	1
carbon	sugar ⟶	cellulose, starch	3	16
oxygen	base ⟶	nucleic acid (DNA,RNA)	88	76
nitrogen	amino acid ⟶	proteins	0.3	3
		Total	92.3	96

Using radiant solar energy, plants convert carbon dioxide and water into different organic compounds, like carbohydrates (sugar is one) and oxygen. This conversion process is called *photosynthesis* and is shown as:

$$\text{solar energy} + \text{carbon dioxide} + \text{water} \underset{\text{respiration}}{\overset{\text{photosynthesis}}{\rightleftharpoons}} \text{carbohydrate} + \text{oxygen}$$

$$CO_2 \qquad H_2O \qquad\qquad (CH_2O)_n \qquad O_2$$

(The designation $(CH_2O)_n$ is the general formula for carbohydrates. For example, $C_6H_{12}O_6$ is the chemical formula for the glucose sugar molecule.)

This photosynthetic activity is responsible for all our fossil fuels of gas, oil, and coal.

The special importance of photosynthesis is that plants can thereby manufacture food from simple and abundant compounds in the atmosphere and hydrosphere. Both plants and animals use the food produced by photosynthesis to build tissues and cells, and the energy bound up in the food molecules fuels their many activities. This process of energy release is called *respiration* and is the chemical reverse of the photosynthesis reaction.

In those early days, the oxygen produced during photosynthesis combined chemically with many elements present in the earth's atmosphere, hydrosphere, and crust. As these compounds became oxidized, photosynthetic oxygen accumulated in the atmosphere as free gas. However, oxygen could build up in large amounts only if some of the photosynthetically produced compounds were neither consumed as food nor oxidized. Look at the photosynthesis/respiration reaction again—note that whatever food and oxygen are produced by photosynthesis are used up by the opposite reaction in respiration. Consequently, for the atmosphere to accumulate large amounts of oxygen, some of the carbohydrates must disappear; this way, respiration cannot occur and thus cannot use up the free oxygen produced by photosynthesis. The most obvious way to accomplish this removal would be to bury some of the photosynthetically produced compounds before they are eaten or oxidized—within sediments, whether as concentrated deposits of coal, oil, or natural gas, or widely dispersed in small amounts within sedimentary rocks. And this is what happened.

Although the bulk of atmospheric oxygen is photosynthetic in origin, some fraction has been produced by the photochemical splitting of water vapor molecules in the upper atmosphere; the free hydrogen escapes to space, while the oxygen accumulates. See Table 1–2.

As the atmosphere became more oxygen-rich, the triatomic molecule of oxygen called ozone (O_3) also began to collect in the upper atmosphere. With time, the ozone created a layer that shielded the earth's surface from high-energy ultraviolet radiation, which is extremely lethal to cellular functions in all forms of life. Until this high-energy radiation was screened out, life was restricted to water, where radiation could not penetrate very far.

Cycles that link all organisms

All life on our planet depends on major cycles in which critical substances like water, carbon dioxide, oxygen, and nitrogen are used over and over again. These substances are withdrawn from the shells of air, water, and rock; incorporated into animals and plants; and eventually returned to

Crust, atmosphere, hydrosphere, and biosphere—the four interacting earth shells.

these shells, from which they will once again be withdrawn. These cycles are outlined in Figure 1–7 and their recycling times in Figure 1–8. Within the biosphere, they link all organisms—animal and plant, marine and terrestrial.

Carbon dioxide is added to the atmosphere by volcanic activity, biological respiration, erosion (especially of limestones), and combustion of fossil fuels. It is withdrawn by photosynthesis and lime-secreting animals and plants.

The oxygen reservoir is controlled mainly by photosynthesis and respiration. Water not only is important in the biological cycle of photosynthesis and respiration, but also has its own cycle: water in glaciers, lakes, rivers, seas, and underground travels from the atmosphere to the surface through rain and snow, then returns to the atmosphere through evaporation.

Nitrogen escapes to the atmosphere through volcanoes; bacteria convert it into organic matter, especially proteins. Nitrogen compounds in animal and plant wastes and dead organisms are transformed by microorganisms into ammonia, urea, nitrates, and eventually into gaseous nitrogen that reenters the atmosphere.

Figure 1–7

Highly generalized cycles of major substances within the earth's solid and fluid shells. Although the details of each cycle are rather complex, the important interconnections among the atmosphere, hydrosphere, crust, and biosphere are suggested.

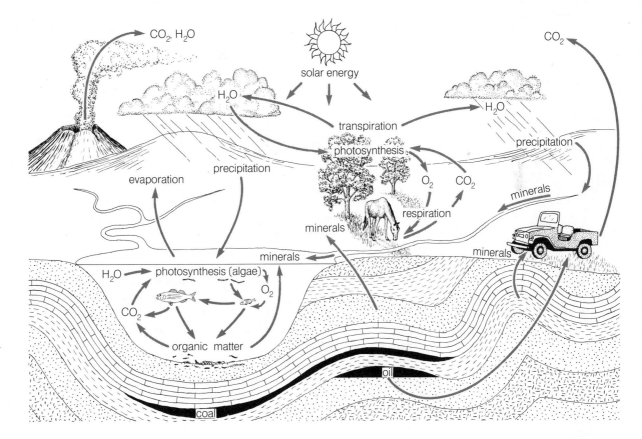

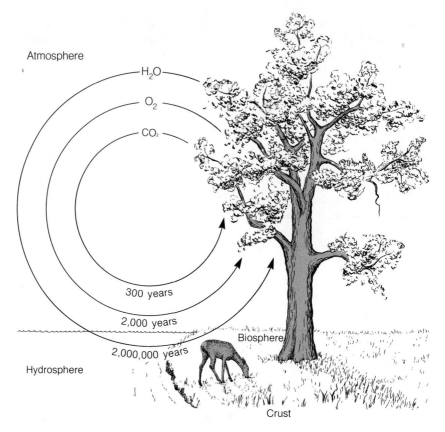

Figure 1–8

Recycling rates of water, oxygen, and carbon-dioxide within the atmosphere, hydrosphere, biosphere, and crust. These residence times depend upon the size of the reservoir of a given substance, and the amount being extracted or added to it annually.

Humans have intervened significantly in these natural cycles of the biosphere. The burning of fossil fuels has greatly increased the return of carbon dioxide to the atmosphere, while the cultivation of legumes, like peas and beans, and the production of artificial fertilizers account for about 10 percent increase of the nitrogen used by natural systems.

What are the sources of energy that keep these cycles in motion, that make the earth a dynamic geologic body and life support system? By far the most significant source of energy is solar radiation, accounting for more than 99.9 percent of all the earth's energy. The remaining 0.1 percent of the energy is divided among the radioactive decay of uranium, thorium, and potassium within the earth's crust; the internal heat stored within the earth during its early formation; and the action of tides. See Figure 1–9. Much of the solar radiation is immediately reflected back into space. Relatively little energy enters the biosphere through plant photosynthesis; some of the energy is used by animals for food, while some is buried in the earth as fossil fuels. Humans use only a small portion of the remaining energy, such as running water for generating electric power, hot water from springs, and natural radioactive elements for medicine.

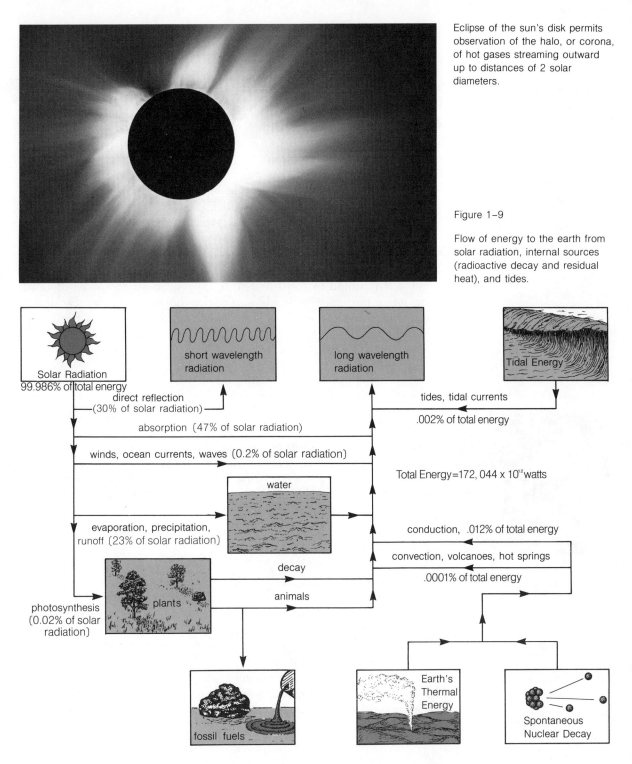

Eclipse of the sun's disk permits observation of the halo, or corona, of hot gases streaming outward up to distances of 2 solar diameters.

Figure 1–9

Flow of energy to the earth from solar radiation, internal sources (radioactive decay and residual heat), and tides.

19

Solar radiation is energy produced by nuclear reactions occurring within the sun in which hydrogen is converted into helium. Only a small portion of solar radiation reaches the earth, and most is radiated back into space by direct reflection from the atmosphere (30 percent) or—after being absorbed by the atmosphere, oceans, and land surfaces—by long-wave heat radiation (47 percent). About 23 percent of the solar radiation drives the water cycle through evaporation, precipitation, surface runoff of water, and melting of snow and ice. A much smaller part (0.2 percent) of the sun's energy drives the circulation of the atmosphere and hydrosphere, resulting largely in what we call climate. The tiniest fraction of solar radiation (0.02 percent) is captured by photosynthesizing plants and is transformed into organic matter which, in turn, supports all the rest of life. Although only an infinitesimal percentage of the total solar energy reaching the earth is consumed directly by organisms, the absolute amount, 40 trillion watts, is enormous.

Some of the energy stored by plants is not consumed immediately as food, but is stored in fossil fuels (oil, gas, and coal) and released much later when burned. We should also notice, however, that a great deal of energy reaching the earth is not used at all. In particular, the energy absorbed by the earth and radiated back into space as heat is virtually wasted.

Viewpoint   **Preston Cloud**

Preston Cloud is Professor of Geological Sciences, University of California, Santa Barbara. Dr. Cloud's particular research interests are the history and evolution of early life. In this Viewpoint he reminds us that the earth is finite and that we humans must learn to live in balance with all its systems.

Striking a Balance with Nature

Any consideration of man's relation to nature should take into account the balanced quality of all natural systems. The concept of a dynamic equilibrium in the flow of matter and energy is as far-reaching as the law of gravity. It applies to all systems throughout the visible universe—from galaxies and stars to the smallest physical, biological, and social units. When any one of these systems is stressed in such a way as to unbalance an existing equilibrium, the system reacts to restore a balance. Ponds and rivers, fields and forests, air masses, species, and even individual animals or organs within species respond in ways that indicate this tendency toward equilibrium.

Homeostasis (the Le Châtelier curve)

The principle of restoration of balance is called the Le Châtelier principle (homeostasis in biology) after its French originator. The records of more than two billion years of biological evolution dramatically illustrate the penalty for species that fail to remain in balance with their changing natural environments: drastic reduction in numbers and eventual extinction. Similar restraints presumably apply even to clever and adaptable *Homo sapiens* in his occupancy of earth and in his use of its resources and ecosystems. In fact, humans are now engaged in a test of the validity of the Le Châtelier principle on a vast scale. As a consequence we are discovering that our task in living at balance with our evolving ecosystem is to understand as best we can how the ecosystem changes, wherein it remains essentially stable, and how our actions may affect the direction and magnitude of future changes.

Consider some examples of the application of Le Châtelier's principle. When earth originated about 4.65 billion years ago, it had little or no atmosphere or surface water. Gravitational compaction, radioactivity, and perhaps other processes raised the earth's internal temperature to the point that volcanic gases poured out and formed an atmosphere. The water vapor contained in these gases condensed to make the primitive ocean. The early atmosphere contained an abundance of carbon monoxide and dioxide but no oxygen. We would consider such an atmosphere to be highly polluted. Then a stress was brought to bear. Primitive algae split the water molecule, thereby producing free oxygen. This process eventually resulted in an atmosphere rich in oxygen and led to the evolution of higher forms of life, including man. If lowly algae could have such a far-reaching effect, even over a long time, it behooves us to consider how nature may respond to the new stresses introduced by humans.

Examples of our effects on and interactions with the land are legion.

Dams on California rivers impede the flow of sand to the sea, while the beach sands move south and eventually to the depths, driven by longshore currents. Result: increased land erosion caused by the sea's relentless attack. Other examples are seen in the replacement of nature's complex ecosystems by man's unbalanced monocultures (farming of a single crop over a large area), which heightens the land's vulnerability to stress. A classic example is the Irish Potato Famine of 1845–1846, in which a million Irish starved and three million emigrated. This massive decrease in numbers established a new balance by reducing the population of Ireland from eight million to four million. We have responded to natural limitations on man's numbers with pesticides, mineral fertilizers, and tougher, more productive seeds. Such achievements, however, have not been entirely an unmixed blessing. Among other results, our techniques for getting the most out of nature have fostered continued population growth in areas that can ill support more people. Asia, for instance, with 85 percent of its potentially arable land already cultivated, is tragically vulnerable to crop failure.

Efforts to expand industrial society have generated other stresses in the form of high rates of consumption of energy and mineral raw materials. Consumption has increased exponentially—like compound interest on a fixed loan. We can roughly estimate the time it takes the consumption rate to double by dividing the rate of increase into 70. World populations are growing 2 percent annually—a doubling time of 35 years. Energy consumption is doubling every 14 years; electricity every 10. Consumption of copper doubles every 16 years, of lead every 25. Even if we could recycle 100 percent of the resources we use, this would provide only half the quantity of any given resource needed in each doubling period. When all of a resource that we can ever recover is only one-fourth gone, the amount left is sufficient for only two doublings. Some of the doubling times have grown shorter as we use ever larger quantities. This is true of our energy consumption. In such instances, warning time between perception of deficiencies and onset of crises becomes shorter, while the potential magnitude of crises becomes larger.

Science and technology can assist in solving some of these problems. We should not let the misuses of past scientific and technological findings cause us to reject the possibility of using such knowledge constructively in the future. Research is needed both to delimit more clearly what the earth's natural limitations are and to understand how closely we can safely approach these limitations as we seek to provide more ample lives to the still underprivileged.

But science and technology cannot do the job alone. We need new economic, legal, and social measures to promote more conserving practices, to reduce inequity, to limit consumption, and to foster nonmaterial enhancements of the human state. Metals have a large multiplier effect on the economy—which is to say that their movement through it generates a flow of dollars many times greater than the dollar value of the basic mineral raw materials. Metallic raw materials represent such a small fraction of the gross national product, however, that there is now little incentive for their more saving use or less wasteful recovery. The price mechanism does not react to conserve most metals. Additions to our capital stock, moreover, come from ever lower grades, requiring movement of ever larger volumes of rock at the cost of ever greater amounts of energy. This aggravates the supply problem and increases the risk of growing environmental degradation. We must enact measures that raise the now unrealistically low cost of nonrenewable raw materials to levels where the price will provide incentive for more conserving use (including efficient recycling) and reduced environmental impact. Depletion quotas (limits placed on the consumption rate of a given resource) and graduated taxes could supplement appropriate increases to the producer to meet the costs of pollution controls and environmental restoration. There are difficulties, but none of them is insurmountable. The mineral resource base is a heritage for all mankind. It should be possible to manage it as such without stifling market incentives.

Unfortunately the oceans are no panacea for these problems. Although oceanic sources will supplement our food and material supplies, they cannot substitute for land sources of many critical commodities. If we increase production by expanding energy input, the environmental and social impacts of energy production will eventually limit our activity, and our conventional energy sources will dwindle as well.

No longer is it reasonable for informed people to speak of infinite resources. Infinity is in the mind and perhaps the stars. Earth is finite. So is its carrying capacity for humans and their industrial society. How shall we react, not knowing exactly what the material limits are? Will we continue to probe until we discover the boundaries, perhaps irreversibly? Or, having the capacity in some sense to foresee the consequences of our actions, will we accept responsibility for them and try soon to strike a new and lasting balance with nature?

Summary

The current physical, chemical, and biological state of the earth results from a long evolutionary development. The earth accreted from a large nebula as a relatively cool, solid body of particulate matter without much of an atmosphere, hydrosphere, or biosphere. Subsequent heating of the early-earth led to its differentiation into a density-layered body with a core of iron-nickel, a mantle of iron-magnesium silicates, and a thin crust enriched in aluminum silicates of the lighter metals.

Volcanic release of the gases bound within the earth's interior formed a primitive atmosphere rich in carbon dioxide, methane, ammonia, and water vapor. With the later appearance of photosynthesizing plants, the atmosphere gradually became enriched in nitrogen and oxygen. Erosion of the lands brought dissolved materials to the oceans, increasing their salt content to its present-day salinity.

All substances flow in cycles through the shells of the earth. Solar radiation generates climates, drives the hydrologic cycle, and supports all life. These internal and external energy sources make the earth a dynamic, ever-changing planetary body.

Thus, seen on a global scale, the earth represents a very small and highly differentiated piece of the primordial galactic cloud. After eons, our planet is like a cosmic onion with layer upon layer of solid, then fluid shells of core, mantle, crust, hydrosphere, and atmosphere. At the boundary between the solid and fluid earth developed the biosphere, whose existence depends upon the interactions among the shells at this boundary. More especially, human survival is closely linked to the living space and resources provided by the crust; the clean air and pure water of the atmosphere and hydrosphere; and the plants and animals living within the intervening biosphere. Our past history is a product of the events that have occurred on and within this planet; our future depends on what it has become and how we manage it.

Glossary

biosphere   The shell of all life that lies between the fluid shells of the atmosphere and hydrosphere and the solid shell of the crust.

core   Central portion of the solid earth composed of iron and nickel. The inner portion is solid, the outer half is liquid.

crust   The outermost solid earth shell composed of rocks rich in sodium- and potassium-aluminum silicates.

dynamic equilibrium   A state of balance within many geologic systems such that, if some conditions of the system change, the balance or equilibrium will shift in a direction to restore the original conditions.

half-life   Amount of time it takes for a radioactive substance to decay to one-half its original amount.

magma   Molten material rich in silicates that crystallizes into igneous rocks of all types.

mantle   Solid earth shell lying between the core and crust, which is rich in iron-magnesium silicates.

oxide   A chemical compound of negatively charged oxygen with a positively charged metallic element.

photosynthesis   Chemical reaction carried out by plants using solar energy to make carbohydrates from carbon dioxide and water.

respiration   Process by which plants and animals release energy bound up in food, producing water and carbon dioxide.

silicate   Chemical compound of two or more oxides, one of which is always silicon dioxide (SiO_2).

Reading Further

Mason, B. 1966. *Principles of Geochemistry*, 3rd ed. New York: John Wiley. A readable and concise account of the history and development of the planet earth. Individual chapters treat the composition of and cycles within the atmosphere, hydrosphere, biosphere, and crust.

Garrels, R. and F. Mackenzie. 1971. *Evolution of Sedimentary Rocks*. New York: Norton. A look at the earth from an integrated chemical viewpoint. Emphasis on the interacting cycles of atmosphere, hydrosphere, and crust throughout geologic time.

The Biosphere. 1970. W. H. Freeman. San Francisco: A collection of essays by leading authorities on various cycles within the biosphere and their relation to the solid and fluid earth.

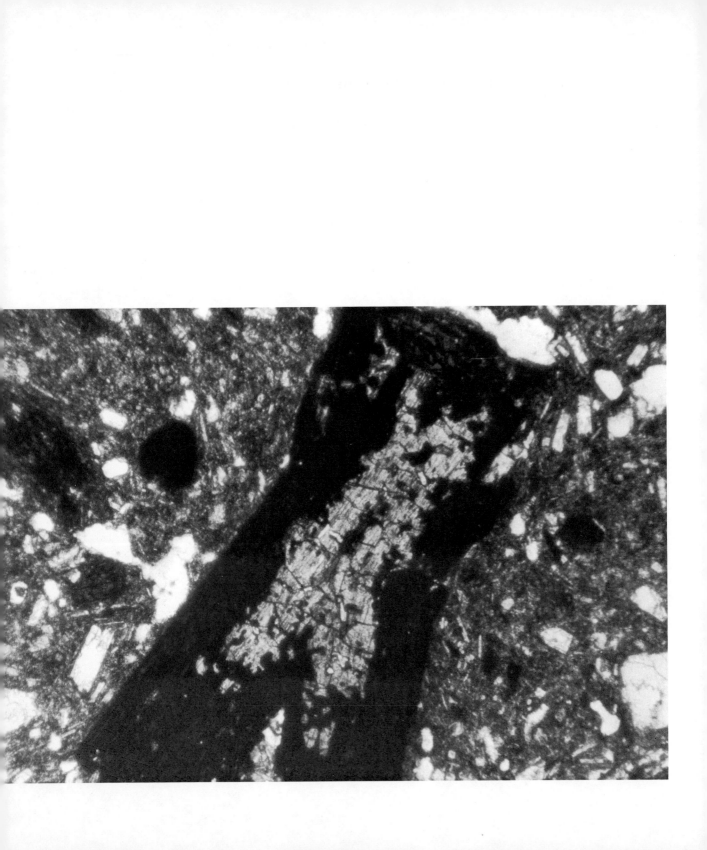

Atoms, Minerals, and Rocks 2

In Chapter 1 we took a bird's eye view of the earth and saw its concentric arrangement of solid and fluid shells, the composition of each shell, and how the shells developed during earth history. In this chapter we want to focus in more detail on the crust, the outermost solid shell, to see what it is made of and how it changes. It is this crustal shell that contains the metals, stone, oil, coal, and soils that provide the resources for civilization. And it is the same shell that receives many of our waste products—whether under the guise of garbage dumps, septic tanks, or deep disposal wells. To understand the resources yielded from the crust and the consequences of accumulating wastes within it, we have to examine the rocks that compose the crust, the minerals that form the rocks, and the chemical elements that combine to make the minerals.

Hot differentiation of the earth produced a crust rich in eight chemical elements: oxygen, silicon, aluminum, iron, calcium, sodium, potassium, and magnesium. In round numbers, these elements account for almost 100 percent of the crust, although many other elements occur in fractions of a percent. All these elements combine to form distinctive solids called *minerals*, each of which is defined by two criteria: what elements are in the mineral and how the elements are geometrically arranged. (It is the repeating geometric arrangement that distinguishes minerals from other solid compounds.) Minerals, therefore, differ from each other in either chemical composition or geometric arrangement of the component atoms.

The rocks within the earth's crust represent a third kind of organization, in which different combinations of minerals produce various kinds of rocks. Thus, as we'll see shortly, the rock called granite has a different assemblage of minerals from the rock called limestone. We classify rocks according to how they and their minerals have formed. *Igneous rocks* solidify from hot molten masses like volcanic lavas; *sedimentary rocks* form from loose sediments like beach sands that accumulate on the earth's

It isn't size that counts so much as the way things are arranged.

E. M. Forster, 1910

surface; *metamorphic rocks* evolve from other rocks that temperature or pressure have altered in composition or internal structure, as, for instance, when a shale metamorphoses into slate.

Many human uses of the earth rely on specific chemical and physical properties of minerals and rocks. The following examples indicate this dependency. Iron for steel comes from the iron-bearing minerals of hematite and magnetite found in some igneous and sedimentary rocks. Oil and gas commonly occur in highly porous limestones and sandstones, two different kinds of sedimentary rocks. Foundations for large buildings constructed on crystalline igneous or metamorphic rocks are more stable than those built on weaker, granular sedimentary rocks. Free-flowing underground water is more prevalent in permeable, coarse-grained sedimentary rocks like sandstone than in impermeable, fine-grained ones like shale.

Thus, to understand and exploit the solid earth shell on which we live, we need to know how rocks originated and of what they are made. Moreover, it is essential to realize that rocks, in turn, are composed of minerals, each with characteristic chemical ingredients and physical structures. Although the earth has overall hundreds of different rocks and thousands of minerals, *most* of the crust contains just a few chemical elements, several kinds of minerals, and a limited variety of rocks. So before we go into the details of the crust, we must first explore its general chemical, mineral, and rock composition.

2-1 Minerals, the Prevailing Few

The major chemical elements of the earth's crust fall into six mineral groups that make up about 95 percent of the rocks in the crust. The remaining 5 percent is composed of a variety of other minerals such as carbonates, sulfides, halides, oxides, epidotes, and zeolites. Notice in Figure 2–1 that silicon and oxygen make up three-quarters of the crust by weight and occur in all these mineral groups. Just which mineral group we'll find in any given part of the crust depends on the chemical elements present and the conditions of pressure and temperature.

The various atoms that compose a mineral join together into a well-defined, repeating, three-dimensional geometric arrangement, or *crystal*. The geometry of these individual crystal structures is as significant in mineral identification as is the chemical composition. For instance, both diamond and graphite are minerals composed of carbon, but their crystal structures are quite different, as you can see in Figure 2–2. In diamond, the carbon atoms are arranged in a repeating unit that is very distinct from graphite. Furthermore, the carbon bonds in diamonds are all of equal strength, whereas in graphite the bonds between the sheets are weaker than those within the sheets. So although they are both composed of carbon, the specific crystalline structure of each differentiates diamond from graphite. In fact, the two crystal structures result in strikingly dissimilar physical properties. Diamond is very hard and transparent, and serves as a cutting tool, an abrasive, or jewelry; graphite is very soft and opaque, and is used as a lubricant. The crystal form of diamond is octahedral (eight-sided); that of graphite consists of hexagonal (six-sided) plates.

Crustal Elements (weight percent) and Water

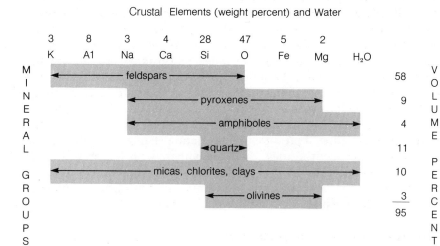

Figure 2-1

Eight crustal elements in the earth's crust are organized into six major mineral groups that account for 95 percent by volume of all the rocks in the crust.

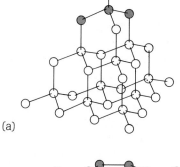

Figure 2-2

Different repeating, atomic arrangement defines diamond and graphite. Carbon atoms arranged like children's jacks form the crystalline structure of diamond (a) which often takes the form of octahedral crystals (b). Graphite's carbon atoms are arranged in interconnecting flat hexagons (c) and expressed in sheetlike, easily cleaved crystals (d).

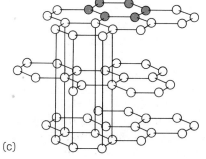

29

A mineral's atomic arrangement is determined by the mutual electrical attraction of the atoms, their size, and their packing within the structure. In turn, crystal structure determines a mineral's hardness, density, and tendency to fracture, or cleave. Not all natural solids are minerals. Some, like coal, amber, and obsidian, are referred to as *mineraloids* because they do not have a well-defined repeating atomic arrangement and therefore are not, strictly speaking, minerals. (Liquids and gases also lack a rigid and well-ordered crystalline structure and hence are not minerals either.)

If pressure and temperature are always the same, but the chemical elements vary in abundance, the minerals that form will differ. As an example, within the mineral group called feldspars, when pressure and temperature are constant, the minerals anorthite or albite will crystallize according to the relative abundances of calcium and sodium, respectively.

Minerals therefore represent solid phases (as distinct from liquids or gases) of chemical elements in equilibrium with local pressure and temperature within the crust. If pressure or temperature changes, or if the relative abundance of elements varies, then the minerals no longer will be in physical or chemical equilibrium. Under a new set of conditions, a different mineral or assemblage of minerals will form to establish equilibrium with the changes in pressure, temperature, or chemistry.

See Skinner's Viewpoint at the end of the chapter for further discussion of atoms and minerals.

The actual time required for these changes may be very long—thousands to millions of years.

Silicates, a key to group character

More than 95 percent of the volume of the earth's crust consists of six mineral groups containing silicon and oxygen; differences among the groups arise from variations in the arrangement of silicon and oxygen atoms within each group. Although the chemical composition among the groups and within a group may vary, it is the silicate structure that principally characterizes each group. For instance, even though feldspars differ chemically among themselves (being rich in either sodium, calcium, or potassium), their common silicate structure unites them as feldspars.

The basic silicate structure is the tetrahedron, in which one small, positively charged silicon atom is surrounded by four larger, negatively charged oxygen atoms. This arrangement has the shape of a pyramid structure, or tetrahedron, with the silicon atom in the center and each of the oxygen atoms at one corner of the pyramid (Figure 2–3). A pyramid formed by four tennis balls with a marble in the center is a good model of the silicate tetrahedron, although magnified about 230 million times!

The ways in which silicate tetrahedra link together define each of the major silicate mineral groups. As you can see in Figure 2–4, the tetrahedra join when each of the silicon atoms shares one or more oxygen atoms with adjacent tetrahedra. *Single chains* of tetrahedra characterize the pyroxenes; *double chains*, the amphiboles; hexagonal *sheets*, the micas, clays, and chlorites; and three-dimensional *frameworks*, the feldspars and quartz. Olivines and garnets have *single tetrahedra* that are joined by atoms other than oxygen, such as iron and magnesium. Each silicate

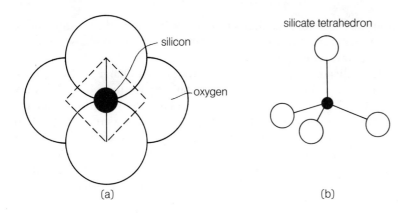

silicon

oxygen

(a)

silicate tetrahedron

(b)

Figure 2–3

The basic building block of the silicate minerals is the silicate tetrahedron composed of a single, small, positively charged atom of silicon surrounded by four, larger, negatively charged oxygen atoms (a); each oxygen atom occupies one corner of the pyramid-shaped tetrahedron (b). Part (a) is drawn to scale—although greatly magnified—while (b) shows ''exploded'' view of atoms to indicate more clearly the silicon tetrahedron.

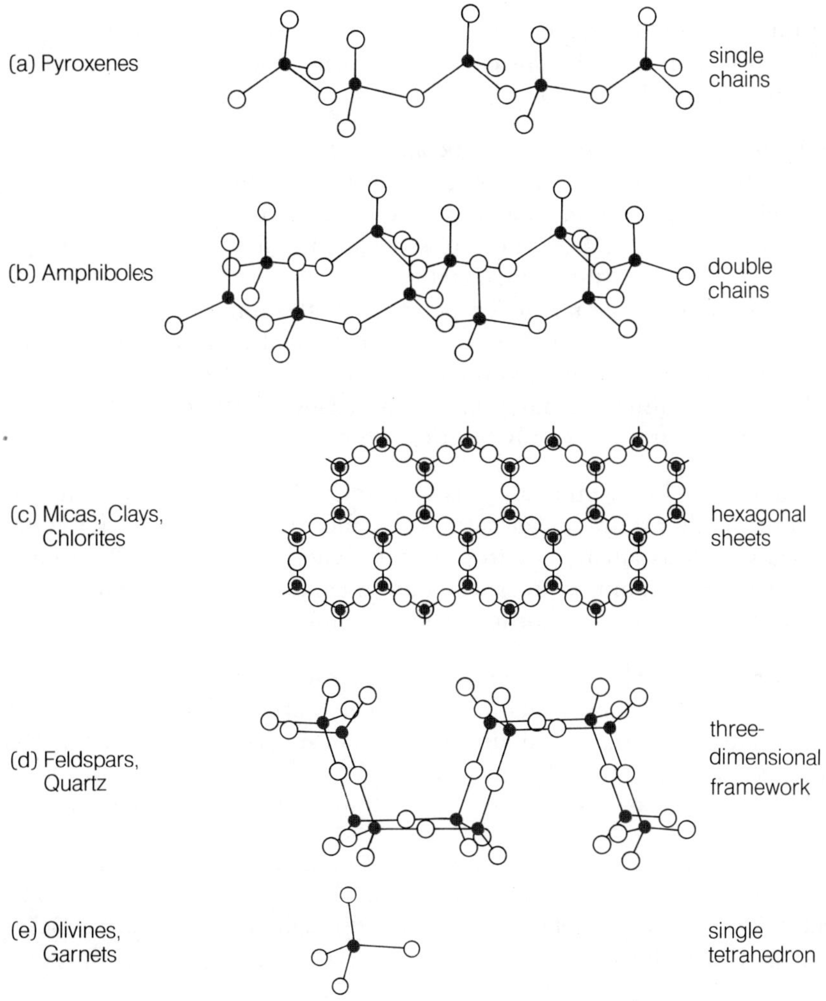

Figure 2–4

By sharing adjacent oxygen atoms, silicate tetrahedra can be linked in a variety of shapes, including single chains (a), double chains (b), hexagonal sheets (c), or three-dimensional frameworks (d). In olivines and garnets, single silicate tetrahedra are joined by atoms other than oxygen, such as iron and magnesium (e).

(a) Pyroxenes — single chains

(b) Amphiboles — double chains

(c) Micas, Clays, Chlorites — hexagonal sheets

(d) Feldspars, Quartz — three-dimensional framework

(e) Olivines, Garnets — single tetrahedron

mineral group, therefore, is characterized by its internal arrangement of silicate tetrahedra; within each group, though, specific minerals are determined by chemical composition.

Individual silicate tetrahedra have an overall, net charge of 4 electrons. This is because the 4 oxygen atoms contribute 8 electrons, of which only 4 balance the 4 positive charges of the silicon atom. Various metallic ions—like iron, magnesium, potassium, sodium, and calcium—with their positive charges latch on to the silicate tetrahedra, thereby balancing the extra unsatisfied electrons. In a similar way, linked silicate tetrahedra have unsatisfied negative charges that attract positively charged metallic ions. Thus, the silicate minerals contain silicate tetrahedra together with a variety of metallic atoms.

Silicate mineral groups

Within each of the six major silicate mineral groups shown in Figure 2–1, many common minerals compose typical igneous, sedimentary, and metamorphic rocks.

Feldspars.   These are framework aluminum silicates in which aluminum atoms may substitute for some silicon atoms in the silicate tetrahedra. Orthoclase is a potassium feldspar, whereas plagioclase is a sodium-calcium feldspar in which the proportions of sodium and calcium atoms can range from pure sodium (the mineral albite) to pure calcium (the mineral anorthite). (Orthoclase provides much of the potassium necessary for fertile soil.) Feldspars are used widely in the ceramic and pottery industry, and most aluminum ores have developed from chemical decomposition of feldspars. This mineral group is common in most crustal rocks—especially igneous rocks—and eventually weathers into clays.

Pyroxenes.   These minerals consist of single-chain silicates with varying amounts of aluminum, magnesium, iron, and calcium atoms. They form minerals such as augite, enstatite, hypersthene, and diopside. Pyroxenes are particularly common in igneous and high-temperature metamorphic rocks; they weather to chlorites upon exposure to air and water.

Amphiboles.   These are chemically and structurally analogous to pyroxenes, except that they contain water (hence are termed *hydrous*) and the silicate tetrahedra have double rather than single chains. Some notable amphiboles are hornblende, tremolite, and actinolite—all found in many metamorphic rocks. Like pyroxenes, amphiboles weather to chlorites.

Quartz.   Like the feldspars, quartz is a framework silicate, but unlike the feldspars, it contains no other chemical elements. Quartz is chemically and physically very stable and is found throughout most crustal rocks. It is used in making glass, semi-precious jewelry, lenses, and sandpaper.

Micas, chlorites, clays.   All of these sheet silicates are structurally and chemically complex. In micas, potassium atoms play a large role in holding the layered sheets together, whereas in chlorites this binding function is taken over by water, iron, magnesium, or aluminum. White or silvery mica is the mineral muscovite, a hydrous potassium-aluminum silicate; while the dark, almost black mica is the mineral biotite which also contains some iron and magnesium atoms. Micas and chlorites occur in rocks of all types.

Clay minerals are hydrous aluminum silicates, with varying amounts of magnesium, sodium, or potassium atoms sometimes replacing part of the aluminum atoms. The mineral kaolinite is a pure, hydrous aluminum silicate. Clays are common in most sedimentary rocks; shales, for example, are composed mainly of clay.

Olivines and garnets.   These are individual silicate tetrahedra joined by iron, magnesium, calcium, or aluminum to make a dense and compact structure. Olivines and garnets are typical of certain igneous and metamorphic rocks that form at rather high temperatures. Garnets are used for abrasives and semi-precious jewelry.

Can you find an example of each of these silicate mineral groups near where you live?

Simplicity in the nonsilicates

The six silicate mineral groups account for about 95 percent of the crustal rocks. The remaining 5 percent of the crust is composed of rarer silicates, like epidotes and zeolites (2 percent), and a variety of *nonsilicates* (3 percent). The chemistry and crystal structure of the nonsilicates are much simpler than the silicates. Typical nonsilicates include salt, fool's gold, gypsum, magnetite, and calcite. What unites these different nonsilicates is their common structure of a small metallic atom such as iron, calcium, or sodium, surrounded by and bonded to larger, nonmetallic atoms like oxygen, sulfur, or chlorine, as illustrated in Figure 2–5.

Despite their low abundance and irregular distribution, nonsilicate minerals supply most of our mineral resources. The simpler crystal structure and occasional local abundance of nonsilicates permit their easier extraction and exploitation as compared with the silicates. To give you an example, it is much simpler and therefore less expensive to extract iron from local concentrations of magnetite or hematite, which contain 70 percent iron by weight, than to process, say, ordinary granite for its iron (which has only about 5 percent by weight).

Nonsilicates are created by the same geologic processes that yield silicates. That is, they may crystallize from magmas, be deposited as sediments, or develop during heating and squeezing of preexisting sediments or rocks. When they are found in local abundance, though, this is usually the result of special conditions. For instance, the copper deposits of Arizona were precipitated from copper-rich, hot watery solutions associated with

Figure 2-5

Atomic arrangement of four common nonsilicate minerals. In each a smaller, positively charged metallic ion is surrounded by larger, negatively charged nonmetallic ions.

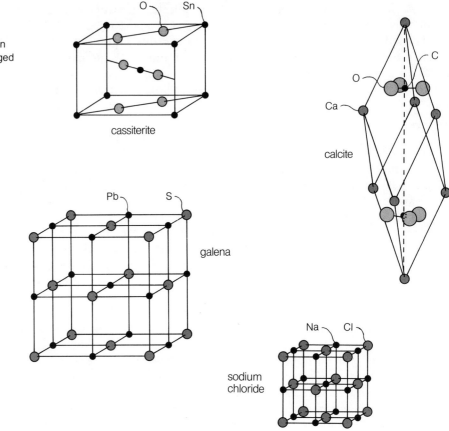

magmas; the salt and gypsum deposits of Michigan were precipitated from a large, evaporating inland sea; and the aluminum deposits of Arkansas are the enriched residue of deep weathering of feldspar-rich igneous rocks.

Nonsilicate mineral groups

The nonsilicate minerals can be conveniently subdivided according to the nonmetallic atoms to which the metallic atoms are bound. The oxides are those in which the metals are combined with oxygen; the sulfides and sulfates in which they combine with sulfur and sulfur oxides; the carbonates in which the metals join with carbonate (CO_3^{2-}); and the chlorides in which the metals combine with chlorine.

Oxides. Oxygen combines with many metals, including iron (the mineral magnetite), iron and titanium (ilmenite), iron and chromium (chromite), copper (cuprite), tin (cassiterite), uranium (uraninite), and manganese (pyrolusite). These metallic oxides may result from magmatic activity and hence are found in small amounts with igneous rocks, but they also occur in sedimentary rocks. For example, hematite, another iron oxide, is a chemical precipitate that forms in aquatic, sedimentary environments. Weathering of soils and rocks produces a common hydrous iron oxide mineral called goethite (sometimes less accurately referred to as "limonite"). Many of these oxides occur in important ore deposits; most are used as pigments in paints.

Sulfides and sulfates.   Sulfur combines with metals such as iron (pyrite), copper (chalcocite), iron and copper (chalcopyrite), lead (galena), zinc (sphalerite), mercury (cinnabar), and molybdenum (molybdenite). Like oxides, these sulfides commonly occur in igneous and sedimentary rocks. The chief sulfate mineral is gypsum, a hydrous calcium sulfate that precipitates from evaporated seawater, and from which plaster is made. Copper, lead, and zinc sulfides are especially important sources for these metals.

Carbonates.   These are minerals whose iron, calcium, and magnesium atoms are bound with carbon and oxygen atoms, and they are most often precipitated from seawater. The iron and magnesium carbonates (siderite and magnesite) are less common than the calcium carbonate minerals (calcite and aragonite). In the mineral dolomite, some of the calcium is substituted by magnesium and yields a calcium-magnesium carbonate. Some organisms like algae and shellfish secrete skeletons of calcite and aragonite, which become fossils in many sedimentary rocks. Calcite is also deposited by groundwater within buried sediments, cementing them into rock; it is an important ingredient in cement and lime used for agriculture. Magnesite is used as a refractory, or heat-resisting compound, in industry.

Chlorides.   When seawater evaporates in arid climates, several minerals are precipitated, especially sodium chloride (halite) and potassium chloride (sylvite). They usually occur together with gypsum, calcite, and dolomite in marine sedimentary rocks. Halite is an important source of table salt. The chlorides, carbonates, and sulfates that precipitate from evaporated seawater are often referred to as the *evaporite minerals.*

And are there examples of these nonsilicates that occur near you?

Recognizing rocks inside and out

As we noted in the beginning of the chapter, rocks are defined by the earth processes that form them as well as by the particular minerals they contain. Igneous rocks crystallize from hot, liquid magmas that originate deep in the earth's mantle and crust. The thick accumulations of lava in the Hawaiian Islands exemplify this kind of formation. Sedimentary rocks are granular aggregates of weathered and eroded rocks that have been scattered over the earth's surface by wind, ice, or water. Fossil-bearing shale beds in the Midwest are solidified marine muds that once covered this region; these beds are now sedimentary rocks. Metamorphic rocks recrystallize from older, preexisting rocks that were heated and pressurized within the earth's crust or were altered by hot solutions flowing through them. The slates and marbles of Vermont are the metamorphosed products of older shales and limestones.

What kind of rocks are you now located on?

Within each of these three categories, we can further subdivide rocks according to their chemical composition and the size of their component minerals. Moreover, the internal structure of grains within a rock may help to identify the rock-forming process. For instance, escaping gases may leave bubbles in lava; shale may contain ripple marks; slate will have its platy minerals parallel to one another, thereby giving the rock a good cleavage. Hence, the presence of bubbles, ripple marks, or cleavage within a rock tells us whether igneous, sedimentary, or metamorphic processes formed it. Throughout the rest of this chapter we discuss each of these major rock types and the processes that created them.

2-2   Igneous Rocks and How They Form

The pressures and temperatures within the earth maintain most of the earth's materials in a solid crystalline state. But two exceptions stand out: the outer core and upper mantle. These two regions exist in a partially liquid state because the prevailing temperature within the earth is hotter than the melting point of the material located there, as you can see in Figure 2–6. The implications of this situation are twofold. First, as we

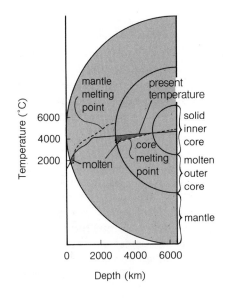

Figure 2–6

The temperature of the earth increases with depth (solid line) and is shown with the melting point curves for rocks composing the mantle and core. Present earth temperatures in the upper mantle and outer core are higher than the melting point of the rocks and hence are in a molten rather than a solid state. The molten upper mantle is the source of many igneous rocks; the molten outer core is the source of the earth's magnetic field.

(a)

(b)

Figure 2-7

(a) Volcanic eruption in the Hawaiian Islands with molten magma extruding through the surface and flowing down to the sea. (b) Extrusive lavas that have cooled and crystallized into basaltic, igneous rocks within the Haleakala crater, Hawaiian Islands. Notice that there have been a number of lava flows and small volcanoes within the crater.

discussed in Section 1-1, the earth's rotation moves the liquid in the outer core, generating the earth's magnetic field. Second, the liquid upper mantle frequently moves in the form of magma into higher parts of the earth's crust and in these cooler areas crystallizes into solid rock. Sometimes these magmas break through the crust and reach the surface as lavas, or *extrusive rocks* like those pictured in Figure 2-7. Often the magmas do not reach the surface, but instead cool and crystallize as *intrusive rocks* within the crust, which you can see in Figure 2-8.

Extrusive magmas tend to cool rapidly. Consequently, the minerals have only a relatively short time to grow crystals and are usually small or fine-grained. Intrusive magmas, on the other hand, are insulated from the much cooler surface temperatures and have more time to grow larger crystals; thus their minerals have larger or coarser grains. When magmas reach the earth's surface, they rapidly release their trapped gases—mainly carbon dioxide and steam—whereas intrusive magmas usually retain their gases. For this reason, extrusions may be quite explosive; volcanoes illustrate such violent magmatic eruptions. The presence of water in intrusive magmas accounts in part for differences in their minerals as compared with extrusive rocks: amphiboles and micas, both hydrous minerals, are more typical of intrusives than extrusives.

Minerals and magmas

Magmas can vary in their chemistry, especially in the relative amounts of silica (silicon and oxygen), aluminum, potassium, sodium, calcium, iron, and magnesium. The mineralogic variations among igneous rocks reflect these original chemical differences within magmas. Thus, igneous

Figure 2–8

Intrusive granitic, igneous rock cutting vertically through fractured gneissic, metamorphic rock in Colorado.

Verify from Figures 1–4, 2–1, and 2–9 that basalt and gabbro are chemically similar to the mantle.

rocks can be classified according to two features: the minerals that compose them and the minerals' average size or texture, as shown in Figure 2–9.

If the mantle is the source of igneous rocks and if, as we noted in Section 1–1, the mantle is relatively homogeneous chemically (rich in iron and magnesium silicates), how can we account for the chemical and mineralogic variety of igneous rocks? We should expect that magmas from the mantle would crystallize into rocks containing abundant iron and magnesium silicate minerals because of the mantle's chemical composition. And indeed igneous rocks like gabbros and basalts are common in the crust, which indicates that magmas from the mantle have penetrated the crust. But there are many other types of igneous rocks within the crust besides gabbro and basalt. From what magmatic sources do they come? In short, there must be a range of chemical variation among magmas to account for the variety of igneous rocks in the earth's crust.

One kind of chemical variation among magmas occurs when rocks and sediments subside or are pushed downward into regions of much higher temperature and pressure within the earth's crust. Such burial happens in regions of mountain building where portions of the crust subside very deeply, bringing along thick accumulations of surface materials. At depths of a few to several tens of kilometers, these rocks and sediments recrystallize, or even melt, and form igneous rocks that are rich in feldspar and quartz.

This process has been called *granitization* because the igneous rocks so formed are granitic in composition and texture. That is, they contain interlocking crystals of potassium feldspar, quartz, and some mica. In this case, then, magmas from the mantle are not the source of these igneous rocks. Rather, these rocks are examples of extreme metamorphism: preexisting rocks and sediments change to rocks with granitic composition and crystalline texture because of large changes in pressure and temperature accompanying deep burial within the earth's crust.

We refer again to granitization in the discussion of metamorphic rocks below.

From simple magma, many rocks

Another mechanism that explains the mineralogic variability among igneous rocks is the progressive change in a magma's chemistry as crystals form and leave it. This process, called *magmatic differentiation*, results in the continual evolution of a magma's chemistry as each generation of crystallizing minerals is removed from it. A magma from the mantle is initially rich in iron and magnesium silicates and, upon cooling, will form gabbros or basalts. But what would happen if the minerals that first began to crystallize—minerals like olivine and calcium-feldspar—were removed somehow from the remaining liquid magma?

From laboratory experiments conducted in the 1920s by N. L. Bowen we know that a liquid magma's chemical composition changes regularly during mineral crystallization, becoming richer in silica, potassium, and sodium. If the early-formed minerals remain in contact with the magma, they react chemically with it. So everything ultimately crystallizes into the typical mineral assemblages of gabbros or basalts. But if somehow the

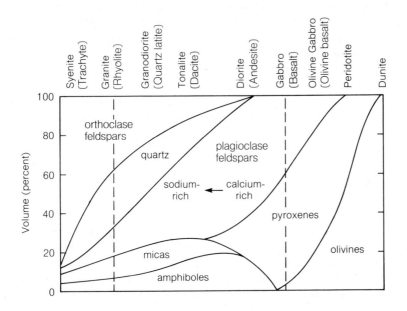

Figure 2-9

Mineralogic variations (volume percent) of common igneous rocks. The names of the coarser grained varieties (usually intrusives) are shown without parentheses, the finer grained types (usually extrusives) within parentheses. Granite, for example, is composed of orthoclase (39 percent), quartz (28 percent), sodium-plagioclase (15 percent), micas (11 percent), and amphiboles (7 percent); whereas basalt contains calcium-plagioclase (41 percent), pyroxenes (57 percent), and olivines (2 percent). To determine the chemistry of these different igneous rocks refer to Figure 2-1.

early-formed minerals are removed from the residual magma, the minerals to crystallize next reflect the larger amounts of silica, potassium, and sodium. With these increased amounts, amphiboles, sodium-rich plagioclases, feldspars, and some mica form. In turn, if *these* minerals are removed, the remaining magma will ultimately crystallize into a rock of granitic composition. That is, the rock will be rich in the quartz, potassium feldspar, and mica from the magma left behind by the earlier-formed minerals. In the photograph on page 26, a thin, translucent slice of andesite—a fine-grained igneous rock—shows a pyroxene crystal (light) altered to an amphibole (dark) through reaction of the pyroxene with the cooling magma.

Field observation of igneous rocks has corroborated Bowen's laboratory experiments and has led to a formal theory of how magmatic differentiation works. Often called Bowen's Reaction Series, the process is outlined in Figure 2–10. During this series of reactions between crystallized minerals and liquid magma, the minerals created from the molten magma reflect its original composition. But, if each generation of newly crystallized minerals is removed from the cooling magma—a removal which obviously prevents the minerals from reacting further with the magma—then each new generation of minerals will be chemically different from any other. The fascinating thing is that such a wide variety of igneous rocks, ranging from basalt to granite, as you can see in Figure 2–10, may originate from an initially homogeneous basaltic magma.

In reading the description of magmatic differentiation, you might have asked, what sorts of geologic processes would remove early-formed minerals and prevent any further reaction with the magma? Among some igneous rocks, early-formed minerals, which have a greater density than the liquid magma, settle out of the magma. In other situations, the magma is squeezed out of the accumulating mass of early crystals and into surrounding rocks. This process usually happens during mountain building, when an intrusive magma is squeezed or deformed during cooling.

Concentration of ore-bearing minerals may occur during magmatic differentiation. To cite one example, layers rich in iron, chromium, and platinum minerals occur in some South African rocks where early-formed minerals settled out of a large igneous intrusion. In other places, some silver and gold deposits were also concentrated in late-crystallizing, quartz-rich veins running through many igneous rocks.

Igneous structures

The extrusion and intrusion of igneous rocks create several kinds of geologic structures, as shown in Figure 2–11. Surface eruptions of magmas may produce horizontal, stratified *lava flows*. At the point of eruption small particles of magma thrown into the air cool rapidly and accumulate as *ash beds*. The conical structures that we call *volcanoes* include both lava flows and ash beds. Magmatic intrusions that do not erupt onto the

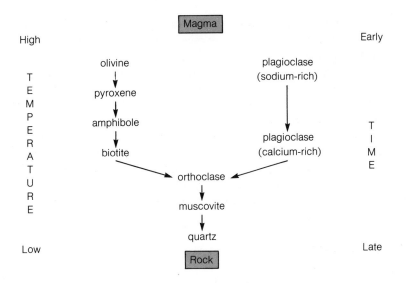

High

T
E
M
P
E
R
A
T
U
R
E

Low

Magma

olivine

↓

pyroxene

↓

amphibole

↓

biotite

plagioclase
(sodium-rich)

↓

plagioclase
(calcium-rich)

orthoclase

↓

muscovite

↓

quartz

Rock

Early

T
I
M
E

Late

Figure 2-10

Mineralogic variations within a cooling magma of mantle-like, or basaltic, composition. If earlier formed minerals are removed from the residual melt, the subsequent rocks will be more granitic in composition. But if the earlier formed minerals are allowed to react continuously with the residual melt, the final mineral assemblages will reflect the magma's original basaltic composition.

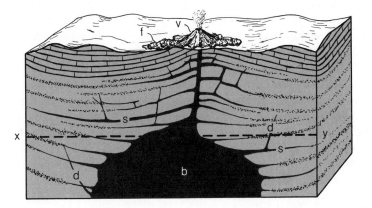

Figure 2-11

This schematic cross section of the earth's crust illustrates some typical structures associated with igneous rock formation. A large intrusive batholith (b) is surrounded by dikes (d) and sills (s) that cut across or lie parallel to the stratification of the rocks being intruded. Some of the magma has penetrated the crust and formed a volcano (v) with lava flows (f). Erosion down to the level of the dashed line (xy) would expose the batholith with its sills and dikes; the rocks in contact with the batholith are metamorphosed by the heat and pressure accompanying the intrusion (see Section 2-5).

earth's surface may inject themselves either as *sills*, if the intrusion lies parallel to the rock beds being intruded, or as *dikes*, if the intrusion cuts across the rocks. Extremely large intrusions, usually found within the central core of mountain ranges, are called *batholiths*. Metal-rich mineral deposits sometimes line the boundary surfaces between these different igneous structures. The deposits were created because the magma had an unusual chemistry or because the hot magma reacted chemically with surrounding rock.

2-3 From Weathering and Erosion, Sedimentary Rocks

As we discussed, minerals crystallize under specific chemical and physical conditions. When rocks composed of those minerals are subject to new

41

physical and chemical conditions, they begin to break down and change. A granite that crystallizes from a hot magma, for example, contains minerals in equilibrium with the temperatures, pressures, and chemistry of the granite's environment. When that granite is exposed to different physical and chemical conditions at the earth's surface, the crystalline mass of minerals begins to disintegrate and dissolve. The granite, in short, is weathered by exposure to air, water, and organisms. The materials produced by this weathering process—which includes rock and mineral fragments as well as dissolved substances—also may be eroded; the materials could be carried away by wind, ice, or running water. The products of weathering and erosion form *sediments*, loose deposits of granular material. Eventually, these sediments may be cemented together to make sedimentary rocks, the second major class of rocks.

Types of sediment

All rocks at the earth's surface are weathered, eroded, and transported, however slowly. Glaciers, running water, and pounding surf gradually wear away the rocks in mountains, valleys, and along coasts. Rock fragments of all sizes—boulders, pebbles, sand, and silt grains—are produced by weathering and erosion, and are reduced in size as they are borne along by winds, streams, and glaciers. Sediments formed near their source tend to mimic the composition of their parent rock. But as weathering, erosion, and distance from the source increase, unstable minerals like those in Figure 2–12 change to chlorites and clays. Moreover, these forces round the rock fragments and sort them by size; finer grained particles separate from the coarser grained ones. Just as a magma can yield various rock types by magmatic differentiation, so too can an igneous rock yield different sediments by the processes of *sedimentary differentiation:* weathering, erosion, and transportation.

Some sedimentary rocks are called *clastic* (from Greek *klastos*, meaning broken) because they originate from surface accumulations of rock debris. The names given to clastic sedimentary rocks depend on the size of their component grains as well as their composition. Eventually, granular sediments like gravels, sands, and muds may be cemented together to make sedimentary rocks like conglomerates, sandstones, and shales (Table 2–1).

Figure 2–12

Role of weathering and transportation in producing sediments. (a) Weathering rates, using the minerals in granite as an example. The iron-magnesium silicates are soon altered to chlorites, while the feldspars gradually decompose into clays. Quartz is strongly resistant to and little affected by weathering. (b) As sediment is transported, sedimentary grains become more rounded and better sorted while the finer-grained clays winnow out.

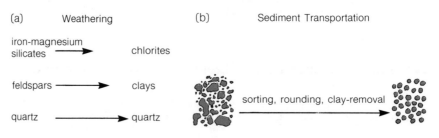

Table 2–1   Origin and Classification of Major Sedimentary Rock Types

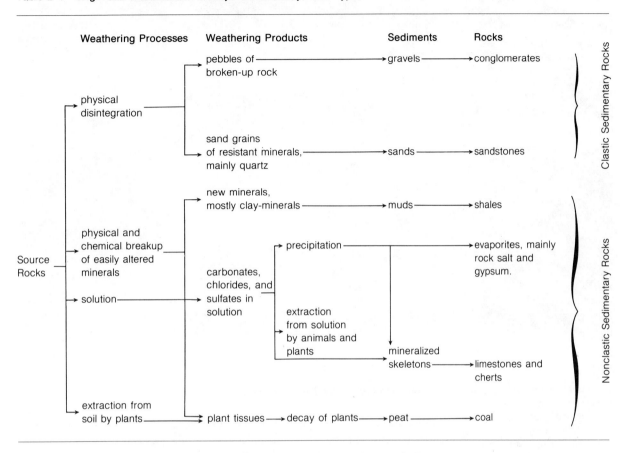

Running water also dissolves rocks and minerals at the earth's surface, transporting the dissolved substances to the sea. *Nonclastic sediments* may form as direct crystalline precipitates from evaporated seawater, such as deposits of rock salt or gypsum seen in Table 2–1. Nonclastic sediments also include deposits left by animals and plants; thick accumulation of dead land plants can make coal, for instance. In the sea, marine organisms extract silica or calcium carbonate for their shells, and their accumulated remains after death produce sedimentary rocks called cherts and limestones, respectively. For a look at some common sedimentary rocks, see Figure 2–13.

When sediments are deposited, many internal structures may form: ripple marks, cross-stratification, graded bedding, mudcracks, and burrows, as shown in Figure 2–14. These structures provide valuable clues as to the surface environment where the sediments formed and that environment's sedimentation processes.

(a)

(c)

(b)

Figure 2-13

Common sedimentary rocks. (a) Conglomerate composed of pebbles and boulders in a matrix of sand particles in upstate New York. Rock accumulated as a gravel along the front of upfaulted mountains almost 200 million years ago. (b) Sandstone of quartz grains in Zion National Park, Utah; notice cross bedding within the rocks. These rocks were deposited as sand dunes some 150 million years ago. (c) Alternating beds of chert and shale in California. The thicker beds are composed of the lithified remains of radiolarians—silica-secreting, single-celled marine animals. Thinner beds are shales that were periodically deposited during the accumulation of the radiolarian cherts.

Hard rocks from loose sediments

Once loose sediments are generated, whether as clastic or nonclastic sediments, how do they become hard rocks? The processes which compact, cement, and recrystallize loose sediments into hard, dense rocks are termed *lithification*. As sedimentary deposits accumulate, the weight of the overlying sediments compresses those below. Water buried within the sediments is squeezed from the compacting sedimentary grains, dissolving some of the smallest ones as it pushes along. The chemical composition of these waters may allow some of the dissolved substances to be precipitated within the grains. The precipitation among the grains gradually cements the whole mass of sediments.

Compacted sediments are cemented in a similar way when surface water—rain and melting snow—percolates down through sediments lying on land. Although the chemical range of possible cements is large, most sedimentary rocks are cemented by either silica (SiO_2) or calcium carbonate ($CaCO_3$). In the finer grained sedimentary rocks, such as shales and limy mudstones, the original tiny clay or calcium carbonate particles recrystallize into a dense, interlocking network of small crystals (Figure 2-15).

Depending on initial grain size and shape, as well as degree of lithification, sedimentary rocks are porous and permeable. *Porosity* is a measure of the open pore space within a rock, and *permeability* relates to the amount of interconnections among the pores. In general, coarser grained rocks are less porous than finer grained ones. For example, about 30 percent of the volume of a beach sand is open pore space, while that of a mud is about twice that. The size of the pores in coarser grained rocks, however, enables fluids within the pores to migrate more freely. Such rocks are therefore more permeable than finer grained rocks whose pores, while more abundant, are much smaller and tend to impede movement of fluids within them. One practical result of this difference in permeability is that many sandstones are good reservoirs of oil or water because their good permeability facilitates the continuous withdrawal of these fluids from the interstitial pores.

Sedimentary strata

Sediments are deposited within the earth's gravitational field by wind and water, and create tabular bodies, or *strata*, lying parallel at the earth's surface. (Notable exceptions are sediments that accumulate as dipping deposits at the foot of mountain slopes or at the base of steep submarine slopes.) Although most sedimentary strata are initially deposited horizontally, later deformation of the earth's crust can distort their orientations into inclined, vertical, or even overturned positions (refer to Section 3–4).

Besides variations in external geometry, sedimentary strata also vary internally in grain size, composition, and structures. That is, sedimentary processes may change laterally across an area receiving sediments, so the resulting deposits vary from one place to another. As an example, gravels may grade into tidal marsh sediments that, in turn, grade into clean beach sands that gradually become muddier offshore. Such horizontal variations in a sedimentary stratum are termed *facies*. We can distinguish facies on the basis of inorganic properties such as grain composition, grain size, and physical structures, as well as biological properties like fossil content and animal burrows.

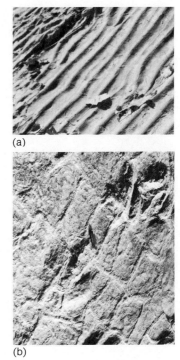

(a)

(b)

Figure 2–14

Some typical sedimentary rock structures. (a) Ripple marks preserved in a marine sandstone that was deposited 125 million years ago in Colorado. (b) Dinosaur footprints preserved in mudcracked freshwater shales in Connecticut. (Which occurred first: the mudcracks or the dinosaur walk?)

2-4 New Rocks from Old

The earth's crust is constantly in motion. Some areas undergo uplift and erosion, while other areas subside and collect sediments. As we have just observed, rocks like granites—and the minerals composing them—form in one geologic environment having specific chemical and physical conditions and later come to other geologic environments with quite different conditions. We therefore might think of sedimentary rocks as the product of conditions at the earth's surface where temperatures and pressures are relatively low and water is abundant.

Figure 2–15

Translucent, thin section of a limestone composed of fossil shells and concentrically precipitated calcareous grains. All the particles are cemented together by clear, fine-grained calcium carbonate deposited within their pores.

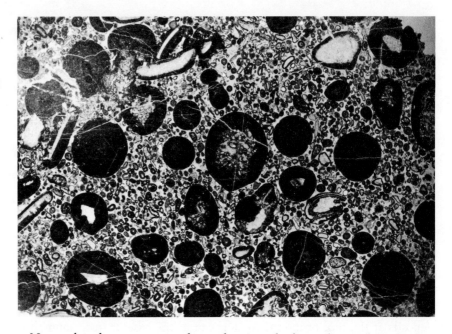

Now what happens to rocks and minerals formed at relatively low temperatures and pressures, such as extrusive igneous and sedimentary rocks, which later experience much higher pressures and temperatures? As you might expect, this third major class of rocks, the metamorphic rocks, is produced from preexisting rocks that are subjected to increased temperature and pressure within the earth's crust. In general, the degree of transformation, or metamorphism, of a rock depends on the chemical and physical differences between the rock's premetamorphic geologic environment and the subsequent metamorphic one. Thus, sedimentary rocks formed at low temperatures and pressures will be metamorphosed more than, say, intrusive igneous rocks crystallized at much higher temperatures and pressures.

Metamorphic processes and structures

Rock metamorphism occurs in two simple ways. First, *local contact metamorphism* happens when nearby increases in temperature, and sometimes pressure, that normally accompany igneous intrusions alter the rocks being intruded. The second method is *regional metamorphism*, whereby rocks are pushed deeply into the earth's crust over a broad region; the resulting rises in temperature and pressure change the rocks. In both, higher pressures and temperatures cause the minerals in the preexisting rocks to recrystallize to new mineral assemblages.

Remembering that minerals are defined by their crystal structure as well as by their chemical composition, and that crystal structure depends on local pressure-temperature conditions, we know that the total chemical composition of rocks being metamorphosed can remain the same while new minerals crystallize. Thus, metamorphic changes in mineralogy do *not* necessarily require accompanying changes in the rocks' overall chemistry. Rather, new minerals crystallize under the new conditions of pressure and temperature. A sandy shale composed of quartz grains and clay particles may metamorphose into a rock called *schist*, in which quartz, mica, and some garnet occur in thin layers. Refer to the thin sections, or translucent slices of rock, in Figure 2-16. What happens, in effect, is that the atoms within the sandy shale crystallize into the higher temperature and pressure minerals of mica and garnet. While this transformation is going on, the silica of the individual quartz grains grows into a mass of crystalline quartz.

As another example of mineral changes during metamorphism, consider the limestone in Figure 2-17 composed of shell fragments scattered in fine-grained mud. When metamorphosed, the limestone converts from a heterogeneous conglomeration of calcium carbonate particles into a *marble*—a coarse and evenly textured interlocking network of calcite crystals. In both examples, the overall chemical composition of sandy shale and schist, and of limestone and marble, remains the same. Metamorphism merely results in the crystallization of new minerals (e.g., mica and garnet) or recrystallization of existing minerals (e.g., quartz and calcite).

It often happens, however, that chemical changes *do* accompany metamorphism. In contact metamorphism, magmatic fluids from an igneous intrusion can penetrate the surrounding rocks and introduce new chemical elements. Or in regional metamorphism, fluids contained in the buried rocks—either groundwater or fluids melted out of the rocks themselves—can migrate into other parts of the metamorphosing pile and create new minerals.

When new minerals recrystallize during metamorphism, they produce numerous structures within the resulting metamorphic rocks. The geometric orientation of different metamorphic structures relates to the stresses

Figure 2-16

Drawings of thin, transparent slices of rock illustrating metamorphism. Sandy shale (a) composed of round quartz grains in a matrix of platy, randomly oriented clay particles is metamorphosed into a schist (b) containing alternating layers of crystalline quartz (Q) and platy micas (M) with scattered garnets (G).

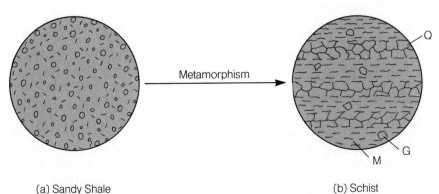

(a) Sandy Shale   Metamorphism   (b) Schist

Figure 2–17

Thin section drawings illustrating metamorphism. Limestone (a) composed of fossil shell fragments in a fine-grained calcium carbonate, or calcareous, matrix has metamorphosed into a marble (b), in which the various calcareous grains have recrystallized to a coarse-textured, interlocking network of calcite crystals.

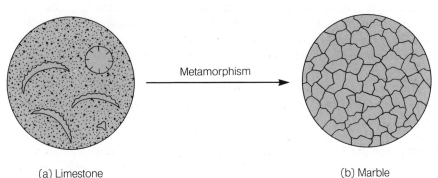

(a) Limestone

Metamorphism

(b) Marble

accompanying metamorphism during the local igneous intrusion or regional burial. In particular, the new minerals grow in directions at right angles to the maximum stress exerted on the metamorphosing rocks. For instance, a parallel alignment of minerals, called *foliation*, develops when thin platy minerals like chlorite and mica recrystallize parallel to each other. Foliation gives the resulting rock a laminated structure, so planes of weakness exist and the rock comes apart, or cleaves, easily [Figure 2–18(a)].

Chemical variations in the preexisting rock may produce minerals that are not only aligned in parallel but also segregated into individual layers. This structure is called *schistosity* and is characteristic of schists. *Gneissic* structures, pictured in Figure 2–18(b), are those in which the newly formed minerals occur in thicker, coarse-grained layers with little or no foliation. If elongated or needlelike minerals like some amphiboles and tourmaline develop in metamorphic rocks, they often lie parallel to one another within distinct layers and give the rock a *lineated* appearance.

Some common rock types

Although there are as many different metamorphic rocks as there are rocks that can be metamorphosed, we need only consider a few of the more common ones. Slates are metamorphosed shales whose clay minerals have recrystallized into micas; parallel orientation of the platy micas gives slate its well-developed foliation and cleavage. Schists are shales that have been more highly metamorphosed than slates. The clays in shales recrystallize to micas and often to garnets. If the shale originally contained scattered quartz silt, the schist will also contain layers of crystalline quartz.

Sandstones composed of quartz grains metamorphose into rocks called metaquartzites. In these rocks the original, round, separate grains recrystallize into an interlocking mass of quartz crystals. Limestone, as noted earlier, recrystallizes in a similar manner, so the original separate calcium carbonate particles fuse into a crystalline mass of marble.

Gneisses are highly metamorphosed rocks of granitic composition—for instance, quartz, feldspar, and some mica. But, as you can see from Figure 2–18(b), unlike granite, the individual minerals in gneisses are segregated into bands or layers. Under high pressure and temperature conditions deep in the earth's crust, watery fluids rich in potassium, sodium, aluminum, and silica melt out of the metamorphosing rocks. These fluids travel upward and react chemically with overlying rocks, producing metamorphic rocks similar to granite in overall chemical composition and minerals (quartz, feldspar, and mica). This is the process of partial melting and metamorphism we defined earlier as *granitization*. As the term suggests, here metamorphic processes grade into igneous processes.

2-5   The Rock Cycle

Energy from the earth's interior causes all these changes in rocks and minerals. Whether the energy comes from residual heat left over from the earth's initial formation or from continuing radioactive decay (especially of uranium, thorium, and potassium), this small portion of the earth's total energy budget is an extremely crucial one. Internal energy drives the slow movement of the upper mantle, and this slow convection leads to mountain building and igneous activity, keeping the earth's crust in

Figure 2–18

Some common metamorphic rocks. (a) Slates in Pennsylvania quarry showing good cleavage parallel to the axis of the upfolded rocks. Note how lighter colored layers within the slate define the shape of the fold. (b) Intricately folded feldspar layers within a gneiss in Alaska. The high pressures and temperatures accompanying metamorphism allowed the rock to flow like a viscous liquid.

(a)

(b)

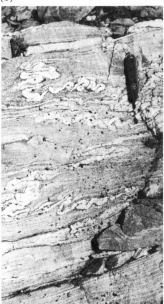

a constant state of flux. Combined with the solar energy that drives the hydrologic cycle and circulates the atmosphere and hydrosphere, such internal energy makes the earth a dynamic, ever-changing geologic body. Figure 2–19 details this continual process of geologic birth and renewal.

So rocks are continually formed, destroyed, and formed again. Heat from the earth's interior melts rocks in the upper mantle and crust and creates magmas. These molten silicates crystallize as igneous rocks like granites and basalts; gases within the magmas seep into the atmosphere and hydrosphere. Weathering and erosion of all three kinds of rocks at the earth's surface by running water produce sediments like sands and muds that eventually reach the oceans. Weathered and eroded rocks not only disintegrate physically into loose, granular materials, they change chemically.

The constant recycling of rocks explains the large discrepancy between the age of the earth (about 4.5 billion years) and the oldest rocks dated so far (about 3.7 billion years).

Figure 2–19

The rock cycle. This flow of earth materials within the mantle, crust, hydrosphere, and atmosphere continually forms igneous, sedimentary, and metamorphic rocks.

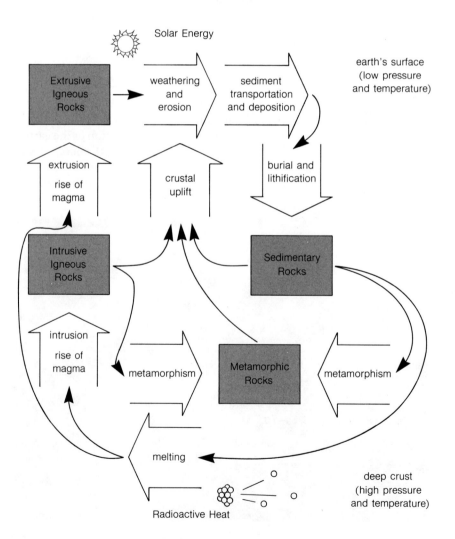

Some sediments form directly in the oceans from the accumulation of the limy remains of algae and shellfish containing calcium carbonate. In shallow parts of the oceans, seawater may evaporate and leave salt deposits. Upon burial, marine sediments (sands, muds, limy deposits, and evaporites) are compacted, cemented, and recrystallized into sedimentary rocks. If the sediments are buried deep enough, the high temperatures and pressures will transform them into metamorphic rocks like slates, marbles, and schists. Partial melting of sedimentary and metamorphic rocks generates new magmas that, upon cooling and crystallization, become igneous rocks. And so the cycle completes itself and continues on.

Vertical movement of materials within the mantle and crust brings buried rocks and magmas to the earth's surface, where they are weathered, eroded, chemically altered, and transported to the oceans. Internal energy sources keep the crust and mantle in motion. Thermal energy from the earth's interior is converted into the *potential energy* (energy of position) of elevated portions on the earth's surface. The uplifted surface of the earth then experiences the effects of weathering, erosion, and sedimentation by wind, water, and ice. The energy for these processes comes from external energy sources, particularly the solar radiation driving the hydrologic cycle. Water vapor in the atmosphere possesses potential energy that converts into *kinetic energy* (energy of motion) in the form of running water which drives erosion and sedimentation. Hence, there is a coupling between the internal energy that lifts mountains and the external energy that wears them down. So long as these two major energy sources keep running, the earth will remain in constant flux.

Viewpoint   Brian J. Skinner

Brian J. Skinner is Professor of Geology and Geophysics, Yale University. Dr. Skinner has done research in geochemistry and mineralogy with special regard to the origin of ores. In this Viewpoint Dr. Skinner points out that while minerals are the common stuff of which the earth is made, economic mineral deposits are few and far between.

Minerals, the Staff of Life

The word "mineral" means different things to different people. To some it means the chemical elements, such as iron and calcium, that are ingested by eating and required by the healthy body. Others use the word when referring to solids in the body such as bones and kidney stones; to still others the word connotes the rich materials dug from mines. This inter-

esting diversity of meaning reflects something that many of us tend to overlook—minerals, in one way or another, are involved in every phase of our existence and well-being. Minerals, not bread, are the real staff of life.

Minerals are naturally occurring, crystalline, inorganic chemical compounds and we find them wherever we look. Soils are mixtures of minerals, but so too are rocks, glaciers, beach sands, and potters' clays. They are the stuff of which the earth is made. Because oxygen and silicon are the two most abundant chemical elements in the crusts and mantles of the earth, the moon, Mars and other nearby planets, most of the minerals in our region of the solar system, are characterized by rigid, three-dimensional linkings of silicon and oxygen atoms. The number of different linkings is large, but we call all such minerals silicates.

Other kinds of atoms (88 sorts appear on the earth) are contained in regularly spaced holes in the silicate structures. Once locked in, they can be released only with great difficulty. To extract most atoms from silicate minerals, the entire mineral structure must be disrupted by breaking the chemical bonds that hold it together. Fortunately, nature has several ways of making available the many elements needed by plants and animals. For example, rainwater falling on a rock starts a slow chemical reaction that breaks down mineral structures and releases elements such as potassium and sodium. These elements can then be taken up by the questing roots of growing plants. This is the process of chemical weathering: without this process most land plants could not survive. But the amount of energy needed for breakdown is enormous (in the example given above, the energy comes, by means of rainfall, from the sun). Indeed, the amount of energy necessary to break down common silicate minerals is so large that we cannot reasonably expect to use these minerals as sources of precious metals, or as fertilizers, or for the myriad other mineral products needed by our complex civilization.

We must therefore seek those few rare and special circumstances where the required elements, such as copper, gold, lead, potassium, and phosphorus, are concentrated into minerals that are not silicates and that can be easily mined and processed. Most local concentrations of valuable minerals, colloquially called ores, form when exposed to a watery solution. For example, the chloride, sulfate, and carbonate compounds dissolved in lake or seawater will become concentrated if the water evaporates. This is the origin of the salt we eat, the potassium minerals with which we fertilize plants, the gypsum used to plaster the walls of our homes, and much of the iron used in our automobiles. In another way, rainwater may remove unwanted material from a solution and leave an insoluble but valuable residue. This is how the ores from which we extract aluminum originate. But by far the most unusual and unique way in which local concentrations appear is through the agency of waters circulating slowly but deeply into the crust. These waters are heated by igneous intrusions and they react with and leach metals from the rocks they pass through. As the circulating solutions travel toward the surface again, they redeposit the leached metals as sulfides and oxides in the veins and porous channels through which they move. The chemistries of both the leaching and deposition processes are complex and like all complex processes, conditions must be perfect in order for an ore deposit to form. The chances that a particular environment will have the necessary qualities are less than one in 10,000. As a consequence, valuable mineral deposits are rare and small. Many experts feel that we are acting in a foolhardy and even irresponsible fashion by mining these rare deposits and using their meager wealth so rapidly.

We cannot draw comfort from the idea that the earth is a cornucopia of mineral riches and that when the crop from the earth has been reaped, the moon, Mars, Venus, and Mercury will be ready to satisfy our appetites. Both ideas are false, foolish, and dangerously misleading. Mineral deposits on earth are few and new deposits are hard to find. In some areas we may already have located all of the accessible minerals. In the area of Europe occupied and mined by the Romans, for example, no new mineral deposits of lead, tin, copper, gold, silver, or mercury have been discovered since the fall of the Roman Empire. Our success in finding valuable minerals on the earth has really been a measure of our success at finding new lands to prospect. But now all of the earth's lands are known, and prospecting for their mineral deposits is far advanced. Days of shortages cannot be far ahead.

The earth's sister planets will not be able to help us out. The more we learn about other planets, the more apparent it becomes that they do not now have, nor probably have they ever had, a hydrosphere like the earth's. The lack of a hydrosphere decreases the other planets' capacity to selectively concentrate valuable minerals by watery solutions. This is not to say that the other planets are entirely barren. A few rare mineral concentrations can form even without watery solutions. But a paucity of water means that weathering and sedimentation, and by analogy the rich diversity of minerals known on the earth, will not be found on the other planets. A paucity of water also means that water solutions circulating in the crust are rare. It is a simple step to conclude that valuable mineral deposits are much less abundant on the sister planets than on the earth. Minerals are the staff of life, but the staff is in danger of weakening and breaking under the mounting pressure of our voracious appetite.

Summary

The earth's solid surface provides the stage upon which the human drama unfolds. We mine it for those materials so crucial to our civilization;

we bury within it the wastes that civilization produces so abundantly. Its soil yields our food and fiber; its topography provides the landscapes for our settlements. Despite the heterogeneity of the earth's surface from place to place, we can describe its overall character rather simply.

The relatively few chemical elements composing most of the earth's crust are organized into six major silicate mineral groups that account for about 95 percent of its volume. Nonsilicate minerals, many of which are important resources, make up another 3 percent of the crust. These silicate and nonsilicate minerals are found in igneous rocks that crystallize from magmas originating deep within the crust and upper mantle, in sedimentary rocks that accumulate at the earth's surface from the weathering and erosion of the crust, and in metamorphic rocks that form from preexisting rocks, which are buried within the crust or intruded by igneous magmas.

Energy from the sun and from the earth's interior keep the rock cycle running continuously, so that igneous, sedimentary, and metamorphic rocks are constantly created anew. The rates at which these processes of mineral and rock formation proceed are very slow by human standards, spanning thousands to millions of years.

This chapter provides us with only a summary of the overall character of the surface of the earth's crust. There are many local, individual exceptions, peculiarities, and geologic "accidents" that we have not discussed yet, but which are of special interest: for example, concentrations of trace elements like gold, chromium, or manganese that form important mineral deposits; local abundances of sand and gravel for construction materials; and peculiar occurrences of contact metamorphic rocks that yield copper or iron.

Glossary

crystal A solid whose constituent atoms have a regularly repeating arrangement.

extrusive rocks Igneous rocks that solidify from magmas that spill out onto the earth's surface.

granitization Formation of igneous rocks of granite composition by the partial melting of preexisting sediments and rocks.

igneous rocks Solids formed from magmas that crystallize within or on the surface of the earth.

intrusive rocks   Igneous rocks that solidify from magmas intruded into the crust below the earth's surface.

lithification   Processes by which loose sediments become hard, coherent rocks, including compaction, recrystallization, and cementation.

magmatic differentiation   Derivation of different igneous rock types from a single parent magma.

metamorphic rocks   Rocks that have new compositions or textures owing to the heating and squeezing of preexisting sediments or rocks. Also includes rocks formed by chemical reaction of preexisting rock with chemically active fluids like magma.

mineral   Naturally occurring solid of specific chemical composition and internal arrangement of constituent atoms.

permeability   Ability of sediments or rocks to transmit fluids within their pore spaces.

porosity   Proportion of pores, or open spaces, within sediments and rocks.

sedimentary differentiation   Processes that produce different types of sediments from a single parent sediment or rock. These processes include weathering, erosion, and transportation.

sedimentary rocks   Solids formed on the earth's surface from sediments that accumulate as loose particles or precipitate directly from solution from natural waters.

Reading Further

Cayeux, A. 1968. *Anatomy of the Earth.* New York: World Universal Library. Introductory treatment of the earth's chemistry and structure followed by a discussion of mountain building and continental drift.

Ernst, W. G. 1969. *Earth Materials.* Englewood Cliffs, N.J.: Prentice-Hall. A short, concise, and chemically oriented discussion of minerals and rocks.

Mason, B. 1966. *Principles of Geochemistry*, 3rd ed. New York: John Wiley.

Turekian, K. 1972. *Chemistry of the Earth.* New York: Holt, Rinehart and Winston. Short paperback that examines the earth from a chemical point of view, including the origin and abundances of the elements, minerals, rocks, radioactivity, and evolution of the planet as a whole and of the crust in particular.

Internal Processes   3

Atoms, minerals, and rocks buttress the earth's outer crustal shell. Yet this crust is in constant flux. Volcanoes erupt, continents inch across the globe, rock strata fold and fracture, mountains grow beneath the sea, and the ocean floor opens. These events point to dynamic forces acting within the crust and upper mantle. The earth's heat produces energy that drives the rise of magmas, carries large parts of the rigid crust and upper mantle here and there, squeezes and deforms rocks, and splits apart the ocean basins. These events do not occur randomly, nor are they distributed haphazardly around the earth. Rather they represent large-scale, intelligible processes at work within the mantle and crust acting over long intervals of time.

Continuing our survey of what the earth is and how it works, we now look into the nature and effects of internal processes because they are intertwined with the way we live. In exploring for minerals, searching for oil and gas, siting nuclear power plants, building homes and factories, and estimating natural hazards, we depend on such knowledge.

In terms of our limited view, the time scale for some of these internal processes seems endless. The rates of crustal uplift and subsidence are so slow that we don't worry about mountains suddenly popping up here or disappearing there. And mineralization, which accompanies igneous and metamorphic activity, takes so long that we cannot count on exploiting such potential resources. But there are always phenomena like earthquakes and volcanoes that occur fast enough to affect humans, and when and where they might occur are questions of considerable concern.

What are the chances of a major earthquake in San Francisco? In Boston? Will downtown Honolulu soon be wiped out by a volcanic eruption or a tidal wave? How long before we run out of copper? Can we find new deposits before then? To respond to such questions we need a broad understanding of what the earth is, how its processes operate,

When mountains fall headlong over hollow places they shut in the air within their cavities, and this air, in order to escape, breaks through the earth, and so produces earthquakes.

Leonardo Da Vinci, c. 1500

and how fast they work. Some answers won't be available even with this background knowledge, but at least we will learn where to look for them. Certain queries lack unambiguous answers, for one reason or another—insufficient information, irregularity of a process, or limitations in scientific technology—so we may not be able to make unequivocal statements or predictions. Sometimes knowing what we don't know, though, is as valuable as being sure of what we do know.

3-1 Plate Tectonics

Which plate do you live on?

In the past decade and a half, scientists have made exciting discoveries about the structure and behavior, or *tectonics*, of the solid earth's outer portion. They have found that the crust and upper mantle are broken into numerous, rigid slabs of rock called *plates*; Figure 3–1 indicates the distribution of the six major plates. Each of these rock slabs is 75–125 km thick, and the major ones are several thousand kilometers wide. Boundaries between plates are traced by midoceanic ridges of basalt, deep sea trenches, folded mountain ranges, or large, vertical faults. The plates encompass both continental and oceanic areas that float as relatively thin, rigid layers on the weaker, underlying mantle. The outermost part of the solid earth, the *lithosphere*, is broken into plates which ride on the *asthenosphere*—weaker, mantle material below them. *Lithos* is from Greek and means stone; *asthenos* is also from Greek and means weak. The lithosphere and asthenosphere differ chiefly in physical qualities: the former is rigid and stronger, the latter, plastic and weaker, as seen in Figure 2–6. The lithosphere/asthenosphere boundary occurs at a 75–125-km depth; this division does not coincide with the crust-mantle boundary found between 8 and 40 km. Moreover, the crust-mantle separation depends mainly on chemical differences, not physical ones; refer to Figure 1–3.

Refer back to Figure 1–3 and compare the chemistry of the core and mantle.

Types of plate boundaries

Boundaries between plates are of three sorts: divergent, convergent, and shear. With *divergent plate boundaries* the plates move away, or diverge, from each other. As they do so, basaltic magmas emerge from the underlying mantle to fill the gap, as you can see in Figure 3–2(a). Magmatic action constantly swells divergent plates along their boundaries. This type of plate boundary lies mostly within the world's ocean basins; an example is the Mid-Atlantic Ridge. Divergent plate boundaries are marked by linear, midoceanic ridges of basalt and frequently experience earthquakes, as do the other two types of plate boundary. The term *seafloor spreading* aptly describes the divergence of plates under the oceans away from midoceanic ridges.

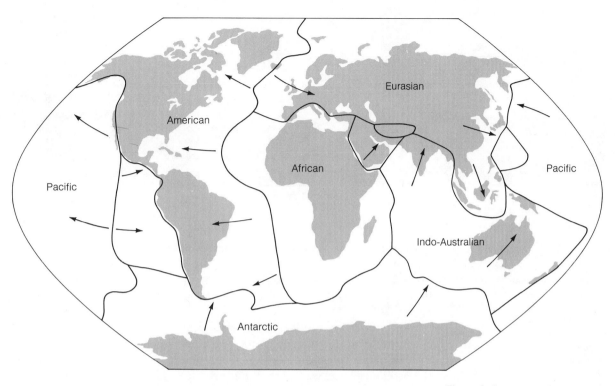

Figure 3–1

Six major plates of the earth: Pacific, American, African, Eurasian, Indo-Australian, and Antarctic. There are also several smaller plates in the East Pacific, Caribbean, Middle East and Central Asia, and Far East. Arrows indicate the general direction of relative plate movement.

Convergent plate boundaries develop when two plates move toward each other and collide. The plane in which they do so is called the *subduction zone.* As one plate hits another, slipping and plunging below it, it is "consumed." The sediments and rocks within the plunging plate are reincorporated into the upper mantle and crust. The collision creates various structures: a submarine trench marks the boundary where one plate plunges below the other; the overriding plate's edge may crumple into mountains and erupt a chain of volcanoes, as in Figure 3–2(b). Both trench and mountains are shaped by one plate dragging down and against another. As the descending plate plunges deep into the mantle, magmas melt out of the rocks and rise upward, spewing forth as volcanoes. This type of plate boundary may lie along edges of continents or within ocean basins—for instance, the Peru Trench and Andes Mountains along the west coast of South America.

The third kind of plate boundary, *shear,* consists of plates that slide past one another along deep, linear, and essentially vertical fractures called *faults;* refer to Figure 3–2(c). With shear boundaries, the plates are neither added to, as with divergent plates, nor consumed, as with convergent plates. Earthquakes occur along the faults but not volcanoes, because no magmas form. The San Andreas fault in California is a shear plate boundary, marking the contact between the Pacific and American plates.

How far is the nearest plate boundary? What kind is it?

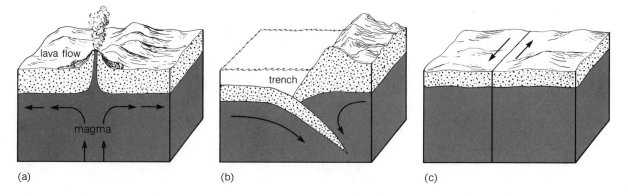

(a)　　　　　　　　　　　(b)　　　　　　　　　　　(c)

Figure 3–2

Simplified views of three types of plate boundaries. Plates diverge (a) with magma coming up from below to form submarine lava flows and midocean ridges. Plates converge (b) and form a submarine trench and mountain system or just a trench as one plate plunges below the other. Plates moving at right angles (c) shear past one another along a deep vertical fault. Igneous activity accompanies plate boundaries of types (a) and (b); in convergent plates (b), the downthrust plate is partially metamorphosed and generates magmas from below.

Perhaps as fast, but not as continuously because plate movement tends to be episodic and irregular.

All three kinds of plate boundaries usually are traced as curved lines on the earth's surface, as seen in Figure 3–1. When mapped in detail, however, these lines are composed of many straight-line segments offset by right-angle, vertical fractures called *transforms*. For a detailed map, see Figure 3–3.

Even though we can locate plates and their boundaries, the mechanism activating plate movement is not well understood. Some earth scientists believe that vertical circulation in the plastic part of the upper mantle, the asthenosphere, transports the overlying rigid plates of the lithosphere across the earth's surface, as indicated in Figure 3–4(a). According to this hypothesis, both lithosphere and asthenosphere move in the same direction. Other earth scientists think that gravity pulls the plates over the plastic asthenosphere, sliding them from the higher elevations of the midocean ridges (swollen by magmas) toward the lower, deep ocean basins. In this case, the lithosphere and asthenosphere move in opposite directions. The latter is illustrated in Figure 3–4(b). Regardless of which interpretation is right, the plates move at a snail's pace. One plate's movement in relation to another is only a few centimeters per year—about as fast as your fingernails grow.

Generating magmas

We have mentioned that magmas emerge along both convergent and divergent plate boundaries; in each instance, magma is generated in a different way. Where plates diverge, material from the mantle below rises as magma and breaks through the ocean floor, producing underwater lava flows. As the plates continue to diverge, basalt accumulates above and below the sea floor, building ridges along the plate boundaries. When plates converge, increases in temperature and pressure metamorphose some of the rocks and sediments within the descending plate. As the plate reaches deeper and hotter portions of the mantle, some of the metamorphosing material melts partially and rises as intrusive and extrusive igneous rocks. Both mechanisms explain why volcanoes are common along midoceanic ridges and parallel to continental margins located inland of submarine trenches; look for this in Figure 3–5.

Magmas that build the midoceanic ridges come from the upper mantle, whose composition is rich in iron and magnesium. Magmas within the subduction zone, from partial melting of crustal sediments and rocks brought down by the descending plate, contain more silica, potassium, and aluminum. Thus, volcanic rocks along midoceanic ridges are basaltic, whereas those along continental margins that border subduction zones are more rhyolitic. The basalts are rich in plagioclase, feldspar, and pyrox-

Verify the mineral composition of basalt and rhyolite from Figure 2–9.

Figure 3–3

Boundaries of the American plate are defined by the midoceanic ridges in the north and south Atlantic Ocean on the east, and by the San Andreas Fault system and East Pacific Ridge on the west.

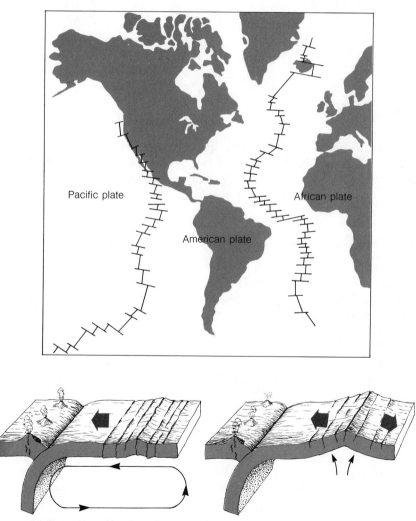

Pacific plate

African plate

American plate

(a) Plates Moved by Convection

(b) Plates Moved by Gravity

Figure 3–4

Two alternate hypotheses that explain plate motion. (a) Shallow convection in the weaker, more plastic asthenosphere carries the overlying rigid lithosphere plates. According to this hypothesis, both the lithosphere and asthenosphere move in the same direction. (b) The lithosphere moves by gravity from the higher elevation of the midocean ridges downward under adjacent plates. In this case, the motions of the lithosphere and asthenosphere are opposite in direction.

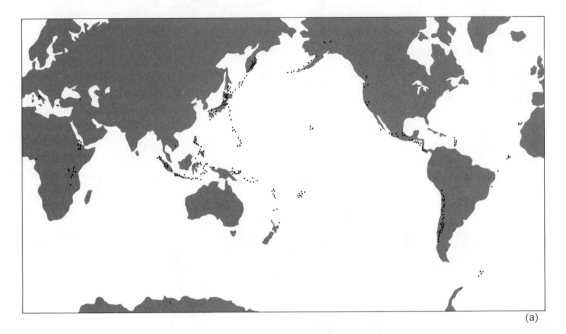

(a)

Figure 3–5

(a) The location of currently active volcanoes around the world. Most volcanic activity occurs along plate boundaries, especially along the contacts of the Pacific and American plates, the African plate with the American and Eurasian plates, and the Pacific plate with the Eurasian and Indo-Australian plates. (b) Earthquake locations for the 7-year period, 1961–1967, from depths of 0 to 700 km. Compare this figure with Figure 3–1, showing major plate boundaries.

ene; while the rhyolites contain quartz, potassium, feldspar, and mica —refer to Figure 2–9. Magmatic activity along plate boundaries appears responsible for some valuable metallic mineral deposits. (See Rona's viewpoint at the end of this chapter.)

3–2   Drifting Continents

If the lithosphere's rigid plates are moving over the weaker, underlying asthenosphere, can we conclude anything about the earth's past geography? Surely the relative positions of continents must have varied during the earth's long history? Well, as a matter of fact, the continents have traveled around, and it was evidence of *past* movement that stimulated discovery of on-going continental motion. The key piece of evidence was the apparently shifting position of the earth's magnetic pole.

Telltale signs

Rotation of the earth slowly moves the outer liquid core, generating a magnetic field. The north and south poles of the magnetic field coincide approximately with the earth's rotational, or geographic, poles. It happens that certain fine-grained minerals—particularly those containing iron—act as tiny magnets and align themselves with the magnetic poles. When igneous rocks crystallize, or sediments compact and lithify, the magnetic

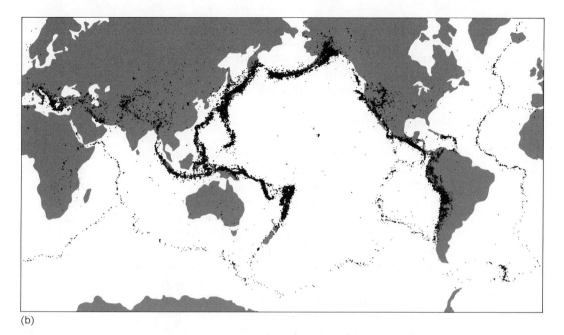

(b)

orientation of fine-grained minerals is "frozen" within them, as pictured in Figure 3–6. Such "freezing" leaves behind *remanent magnetism*. That is, the individual mineral grains act as small compass needles that not only point toward one magnetic pole, but also dip from the local horizon toward the pole as they approach higher latitudes.

A technique for determining remanent magnetism, paleomagnetic measurement, locates the latitude and pole direction of the place where rocks were magnetized before they solidified. Measurements of remanent magnetism have been made on many rocks of diverse ages from all the continents; these calculations have yielded the surprising result that the magnetic poles, and by implication the geographic poles, have wandered throughout

This is roughly analogous to the broken watch or stopped clock in detective stories that provides a clue about the time the crime occurred.

Figure 3–6

Magnetic alignment of fine-grained iron oxide minerals in a basalt sill (a) and in marine magnetite muds (b). Grains in the basalt crystallize in the direction and dip of the local magnetic field. Erosion of the basalt releases the magnetite grains and they are carried to the sea where they are deposited, realigning themselves with the magnetic field. The grains point in the direction of the North Pole and dip from the horizontal in proportion to local latitude.

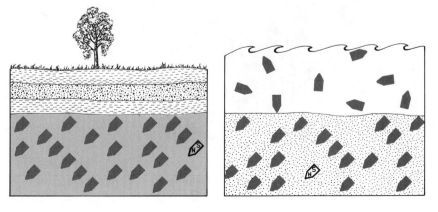

earth history. A related observation is that the paths of polar migration for particular continents become increasingly dissimilar as we trace the paths backward in time; see Figure 3–7. So, knowing that there is only one pair of magnetic, or rotational, poles, we conclude that the continents must have also changed their spatial orientation with respect to each other.

More support for wandering continents

So far, we have used paleomagnetic measurement to establish that continents and geographic poles have wandered during the last several hundred million years. And we have seen that continental drift is related to plate tectonics, in that lithospheric plates move laterally over the plastic asthenosphere, carrying landmasses with them. Paleomagnetic evidence reveals that the continents of the southern hemisphere, in particular, began to fragment from one large "supercontinent" about 180 million years ago, and since then have been drawing farther apart from each other. Looking at Figure 3–8, you can see that South America has moved westward from Africa; Antarctica has traveled southward; and Australia and the Indian subcontinent have drifted northward. The collision of India with the Eurasian landmass helped form the present-day Himalayan mountain system. Going even further back in time, we can find evidence that North America was once joined to Eurasia but has since moved to the west.

Figure 3–7

Paths of migrating poles. The paths diverge as we go further back in time. If the continents were fixed relative to each other, the migrating paths of the poles would be parallel.

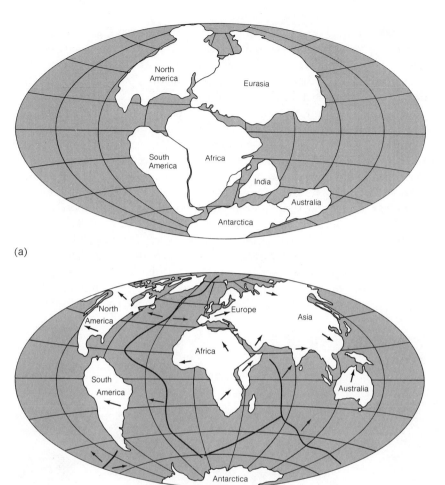

(a)

(b)

Figure 3–8

Fragmentation and dispersal of the Southern Hemisphere continents beginning about 180 million years (a) ago and continuing to 50 million years from now (b).

Other kinds of geologic evidence also support the conclusion that continents have wandered. Remains of fossil plants, ancient coral reefs, and evaporite deposits indicate that some continental areas had quite different climates, and hence different geographic locations, in earlier times. As an example, coal deposits and fossil amphibians and reptiles in the frozen wastes of the Antarctic are convincing indicators of an earlier, more tropical climate for that continent.

3-3 Earthquakes

Those occasional violent shudderings of the earth that we call earthquakes result from sudden movements of rocks within the earth. If we understand the massive movements of plates, it would seem natural and logical that

Figure 3–9

(a) Small scale normal fault in flat-lying sandstones and shales in Colorado Plateau, Utah. Notice displacement of thin, light-colored bed near person's hand owing to rupture of rock strata. (b) Large scale strike-slip fault—the San Andreas—in California. Fault trace is marked by valley through mountains and is the boundary of the Pacific (left) and American plates. The Pacific plate is moving northward relative to the American plate, which is moving southward, as can be seen by displacement of streams that transect the fault.

earthquakes should occur along their boundaries. And as Figure 3–5 shows, principal zones of earthquake activity do coincide with major plate boundaries. Although the plates are under constant stress—pulling apart, colliding, or shearing—friction between plates must be first overcome by the accumulating stresses before they actually move.

Earthquakes also are caused by rupturing rocks; that is, when stresses on rock finally overcome their internal strength, the rocks suddenly fail by breaking, as in Figure 3–9. When rocks are squeezed or pulled apart, they do not immediately crush or break—the squeezing or pulling forces have to build up enough stress to overcome the cohesive forces holding the rocks together. These cohesive forces include the physical bonds among a mineral's atoms and those among the minerals which compose a rock.

Small earthquakes occur virtually everywhere on the earth and are not limited to plate boundaries, although the strongest earthquakes are likely to happen there. In general, earthquakes not associated with plate boundaries are less frequent, weaker, and more irregular in occurrence. They are caused by forces such as volcanic activity, gravitational stress, and rebound of the crust following melting of ancient ice sheets.

(a)

(b)

Focus, epicenter, and seismic energy

The point in the earth's interior where an earthquake begins is called the *focus* of the earthquake; the point on the surface of the earth directly above the focus is called the *epicenter* (Figure 3–10). Earthquakes originate at virtually all depths of the crust and upper mantle. A shallow-focus earthquake is one whose focus lies less than 50 km from the surface. These earthquakes are widespread and usually less severe than deep-focus earthquakes, which occur down to depths of several hundred kilometers. Particularly important sources of deep-focus earthquakes are located along subduction zones; the oceanic plate plunging deeply below the continental plate causes earthquakes in these areas. Such deep-focus earthquakes punctuate the movements of the plates as they jerk past each other.

Energy radiates outward from an earthquake's focus in the form of concentrically expanding waves of vibration. Depending on the initial energy released, the depth of focus, and types of rocks through which the energy moves, these vibrations travel great distances at speeds usually ranging up to 12 km per second (almost thirty thousand miles per hour). Of course, such vibrations at the earth's surface may be strong enough to shake and topple natural and man-made structures.

The energy released by an earthquake passes through rocks in three different ways: as compressional waves, shear waves, and surface waves. *Compressional waves* transmit energy through the back-and-forth motion of rock particles, which move in the same direction as the wave does. With *shear waves*, the energy is transmitted by particles moving at right angles to the direction of the waves, as indicated in Figure 3–11. *Surface waves* move along surface discontinuities within rock strata, or the waves move along the surface of the earth itself. In these waves, the energy is transmitted in an up-and-down motion much like that of ocean waves. All earthquakes send out all three kinds of energy waves; they can be detected by sensitive instruments called *seismographs* even if the waves are not strong enough to be felt by humans. An earthquake's focus and epicenter can be located and its energy can be determined by studying the records, called *seismograms*, recorded by seismographs. Compressional waves are usually referred to as *P* waves (primary) because they travel faster than shear waves, often referred to as *S* waves (secondary).

We measure the energy of an earthquake at its focus by the *Richter Scale*, while we record the intensity of its effects at the earth's surface by the *Modified Mercalli Scale*. The Richter Scale is logarithmic, ranging from 0 to more than 8. (The largest earthquakes recorded so far have measured 8.6 to 8.9 on the Richter Scale.) Each increase of one unit on the Richter Scale corresponds to a tenfold increase in amplitude of seismic waves recorded on seismograms. For example, the 1906 San Francisco earthquake measured 8.3, while the 1971 San Fernando earthquake measured 6.6 on the Richter Scale. In terms of energy released, the San Francisco

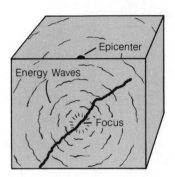

Figure 3–10

The focus of an earthquake, where there is movement along a fault. The earthquake's epicenter is the point on the earth's surface directly above the focus.

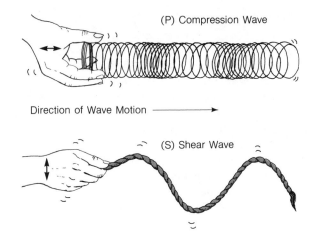

Figure 3–11

Relative motion of the earth as compressional (P) and shear (S) waves from an earthquake pass through it. The motions are like those when a spring coil (a) is jerked forward and backward and when a rope (b) is whipped from side to side.

earthquake was several tens of thousand times as great as the San Fernando earthquake. The Modified Mercalli Scale is divided into 12 degrees of earthquake intensity based on observed surface effects, including reactions of people and animals, damage to buildings, and changes in topography or land forms; it is described in Table 3–1.

3–4 Crustal Deformation

Besides volcanoes and earthquakes, plate tectonics produces another by-product: long, linear mountain chains coinciding with major plate boundaries. For example, the great mountain system of the Americas runs for thousands of kilometers, north to south, from Alaska through Canada and the United States (Rocky Mountains) into central America and down along the western coast of South America (Andes Mountains). A second, related mountain system skirts the western rim of the Pacific Ocean, starting on the Kamchatka Peninsula of northwestern Asia and tracing its course—much of it underwater—through the islands of Japan, Malaysia, New Guinea, and New Zealand. A third, equally extensive mountain system runs west to east, beginning in north Africa (Atlas Mountains), passing through central Europe (the Alps) and the countries around the eastern Mediterranean (Caucasus Mountains), and continuing to the majestic Himalayas of central Asia. These three mountain systems coincide with the boundaries of the converging and colliding plates shown in Figure 3–1. A fourth mountain system consists of a huge network of submarine ridges that runs some 70,000 km through the north and south Atlantic, Indian, and east Pacific Oceans. But unlike the other three continental mountain belts, this submarine chain was created by diverging plates that allowed new crust to form out of magmas rising from below.

68

When looking at a map, we can readily see that other mountains do *not* coincide with plate boundaries. These are topographically high areas left behind by erosion or faulting. Land surface does not erode evenly: some rocks are more resistant to erosion than others and thus they stand higher after a period of erosion than the surrounding weaker rocks. The Ozark Mountains, in the Midwest, for example, stand above the surrounding region because they are mostly composed of resistant granitic rocks. In other areas, like Nevada, rocks are cut by numerous fractures or faults. Some parts of the earth are tilted upward, others downward, along these faults, and erosion of the tilted fault blocks creates mountains and valleys.

Table 3–1   Earthquake Intensity and Magnitude

Modified Mercalli Intensity	Characteristic Effects at Ground Level	Corresponding Richter Magnitude at Earthquake Focus	Ground Acceleration as Percent of Gravity
I	Detected only by seismographs		0.1
II	Felt only by a few people, usually in tall buildings	3.5	
		to	0.3
III	Vibration like that from a passing truck; tall buildings may sway	4.2	
IV	Felt by most people; dishes rattle, objects swing	4.3	0.5
		to	
V	Awakens most people; buildings tremble, windows may crack	4.8	1
VI	Frightens most people; furniture moves, some walls crack	4.9 to 5.4	2.5
VII	Frightens everyone; some damage to poorly built buildings	5.5 to 6.1 Managua, Nicaragua, 1972, 5.6	5
VIII	General alarm, some panic; damage to many buildings	6.2	10
		to San Fernando,	
IX	General panic; ground cracks, most structures damaged	6.9 1971, 6.6	25
X	Many ground fissures, landslides, dams break, pipelines rupture	7 to 7.3	50
XI	Few buildings left, standing bridges down, public service lines out	7.4 to 8.1	75
XII	Damage is total; little left standing; many obvious changes in surface topography	greater than 8.1 San Francisco, 1906, 8.3 Alaska, 1964, 8.5	100

Mountain belts

As we have noted, the three continental mountain chains result from the slow convergence and collision of various plates. During convergence, rocks within the plates fold and become faulted—that is, bent and broken. Some rocks buckle upward as mountains; others push downward into the crust and metamorphose. Still others may granitize and rise upward as intruding batholiths within the overlying folded and faulted rocks. These igneous and metamorphic rocks produce a crystalline core of rocks surrounded by deformed sedimentary rocks as in Figure 3–12. On the continental plate, sedimentary rocks are deformed during orogeny, while those on the oceanic plate are either skimmed off at the trench or carried down with the descending plate and metamorphosed. This process of mountain-building is called *orogeny;* the developing mountain system, or *orogenic belt,* is a long, linear zone of igneous and earthquake activity and deformed rock strata. It is tens of thousands of meters thick and several hundred kilometers wide. Along its flanks lies a mantle of sediments left by erosion of the uplifted rocks.

Figure 3–12

Cross section of a mountain system resulting from the collision of two plates. Rocks on the continental plate are folded and faulted, while those on the oceanic plate are metamorphosed and granitized as they plunge below. Magmas from the granitization intrude overlying rocks as an igneous or granitic batholith; some magmas penetrate the surface and form volcanoes.

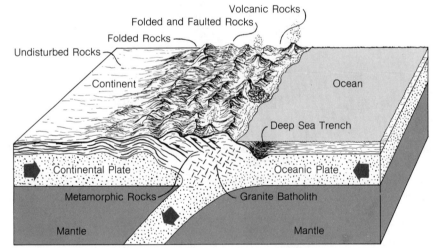

Inactive or quiescent orogenic belts also include the older and deeply eroded mountains of the eastern United States (Appalachians), Scandinavia, British Isles, western Europe, eastern Australia, and elsewhere. Like the younger, more active mountain systems, these older orogenic belts are interpreted as boundaries of earlier plate collisions.

Faults and folds.   As you might deduce, rocks are deformed in two ways during orogeny: breaking or bending. If the stresses accompanying mountain-building are applied rapidly, the rocks in the earth's crust behave as rigid solids and break suddenly when their internal strength is overcome. If stresses act over longer time spans, or if the rocks are near their melting points and thus are less brittle, rocks may react like viscous liquids— although still solid, they yield to the stresses by bending rather than breaking.

Planes along which rocks fracture are either faults, if the rocks on both sides of the break have moved along the fracture, or *joints*, if the broken rocks are not displaced. Fault planes tilt or dip at different angles in the earth, and faults are categorized according to the relative direction of rock movement as well as degree of dip; see Figure 3–13. Faults come in all sizes—from those just a few meters long to the more than 1600-km long San Andreas fault running almost the length of California. Joints are linear breaks that may dip in a number of directions, too, but are of much smaller scale (Figure 3–14).

One meter equals 39.37 inches or 1.1 yards and contains 100 centimeters.

As we noted, rocks may bend rather than break. Sedimentary rocks are often bent into a series of upfolds, called *anticlines*, and downfolds, or *synclines*, as illustrated in Figure 3–15. Metamorphic rocks that have been heated and squeezed also may be folded. Usually these folds are more intricate than those in sedimentary rocks because higher temperatures and pressures allow metamorphic rocks to flow rather than bend; you can see typical results in Figure 3–16.

Unconformities.   Within a pile of sedimentary rocks, often you can see surfaces of erosion called unconformities. For example, after rocks have been uplifted and eroded, they may be covered with sediments. The

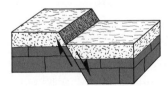

(a) Normal Fault

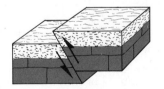

(b) Reverse Fault

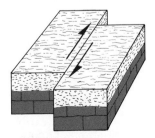

(c) Strike-slip Fault

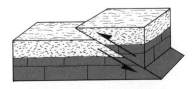

(d) Thrust Fault

Figure 3–13

Four common types of earth faults. Fault planes dip steeply (about 60° from the horizontal) in (a) and (b), but the relative direction of vertical displacement differs. In (c) the relative displacement of the two fault blocks is horizontal along a vertical fault plane. In (d) the fault plane is more shallow than in the other faults, lying about 30° from the horizontal. Here one fault block moves up the fault plane relative to the other fault block.

Figure 3–14

Devils Postpile in California showing well developed vertical joints that formed when basaltic lava cooled and contracted. Individual columns are about 60 centimeters in diameter.

boundary between the older, eroded rocks and the younger, overlying sediments (or sedimentary rocks) records the uplift and erosion event. The word "unconformity" is used because the rocks above and below the erosion surface do not conform in appearance or method of formation. Common unconformities are those found between crystalline or deformed rocks below and flat-lying or gently warped sedimentary rocks above (Figure 3–17).

Although faults, joints, folds, and unconformities exist in many different geologic settings, they are rather common and well-displayed in orogenic belts. For this reason they help earth scientists to determine the orientation and timing of mountain formation. Folds and faults in the Appalachians, for instance, indicate that major mountain-building stresses pushed from the east and southeast to the west and northwest. The age of the rocks and the unconformities within them demonstrate that two large orogenies occurred in the Appalachians: one approximately 450 million years ago, followed by another about 100 million years later.

3-5 Toward an Ideal Balance

Besides plate tectonics and orogeny, another process within the earth's crust keeps it in motion. And like plate movement and mountain-building,

Figure 3–15

Anticline (upfold on left) and syncline (downfold in middle) in deformed sedimentary rocks of the French Alps.

Figure 3–16

Flowage in gneiss. Heating and squeezing during metamorphism have caused the rock to flow like a viscous liquid. Layering results from segregation of various metamorphic minerals.

Figure 3–17

Angular unconformity shown by flat-lying shaly limestone resting on vertically upturned sandstone in Colorado. Hundreds of millions of years separate the deposition, lithification, uplift, and erosion of the sandstone before the limestone was laid down.

it is responsible for some earthquakes and changes in land elevation. This process is the very slow movement of crustal and mantle materials to equalize differences in density within the crust and mantle. The term *isostasy* (from Greek, meaning equal-standing) is applied to the state of balance—never ideally achieved—in which different parts of the crust and mantle maintain their elevation according to their density.

Riding high and low on the mantle

Refer back to Section 1–1 and Figure 1–5.

The thickness of the earth's crust varies from continents to ocean basins as we noted in Chapter 1. Continental crust thickens through mountain-building, and may range up to 65 km by accumulating rocks enriched in less dense minerals like orthoclase feldspars and quartz. In contrast, under the oceans lies only a thin crust, up to 8 km thick, that contains denser minerals like pyroxenes and plagioclase feldspars. Consequently, the greater volume of less dense continental crust sits higher on the underlying mantle than does the less voluminous, denser oceanic crust. And the thicker the continents are, the higher they sit. Lofty continental elevations and low oceanic elevations result from these differences in crustal thickness and density.

Processes that alter crustal thickness and density, such as heating and cooling of rocks, erosion of continents, deposition of sediment, or piling or melting of glacial ice, disturb the crust's isostatic balance. To achieve a new balance, the crust rises or sinks on the underlying mantle.

We can clarify our understanding of isostasy by drawing on a simple analogy. Suppose you fill up your bathtub with water; predictably, the elevation of the water surface is quite even. Now suppose that you throw in some floatable items: plastic ducks, tennis balls, small pieces of wood, a hollow metal sphere, ice cubes, and so on. Two things happen: first, the water surface becomes uneven, dipping under each of the floating objects and rising up around them; second, the tops of the objects sit high or low depending on their density, that is, on their mass per unit volume. The plastic duck, for example, is of quite low density and sits well above the water, whereas the tennis ball sits further down and hardly projects above the water surface at all. In the same way, the ice cubes, wood, and metal sphere stick above and below the water surface in proportion to their density. In short, the less dense materials (the objects you threw in) float on the more dense material (the water), and their heights above and below the water are determined by their densities.

Turning to the earth's crust, we can apply our analogy. When portions of the crust increase or decrease in density, they accordingly sink or rise on the underlying mantle below. The plastic character of the mantle allows this isostatic adjustment by flowing out from under a sinking area or moving in beneath a rising one (as the bath water does when you throw in or remove a floatable object).

Thus, having a tremendous volume of relatively less dense rocks, mountains float on the plastic mantle. They not only sit high above the surrounding topography, but also extend down in the material on which they float (just the way the floating objects in the bathtub stick down into the water) Even though continental lowlands have the same density as mountains, they sit at lower elevations because they have much less volume. And so we can see in Figure 3–18 that the continents with their mountains and lowlands are thickened slabs of lighter rock which float on the denser mantle.

Maintaining isostatic balance

When plates collide, crustal rocks thicken along their margins: sediments on the plates are pushed together and granitic magmas accumulate. (The igneous rocks are relatively enriched in less dense minerals like quartz and feldspars.) Sediments and sedimentary rocks also contribute to the thickening by skimming off the oceanic plate as it plunges below the continental plate; they compress together at the seaward margin of the

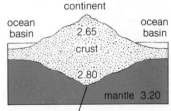

Figure 3–18

Continent of less dense rock floating on denser mantle. Isostasy requires that the less dense the rocks and the thicker they are, the higher they will float on the underlying mantle. The thickened portion of the continent extends upward as mountains and downward as mountain roots.

continental plate as outlined in Figure 3–12. As this wedge of less dense material thickens, it rises upward, floating higher and higher on the underlying mantle.

The high topographic relief of many mountain chains is caused by an enormous increase in volume of less dense rocks during orogeny. But when orogeny ends, the high-standing mountains erode and begin to diminish in volume and elevation. However, mountains, like icebergs, extend part of their mass below the surface and they rise upward to maintain their isostatic balance (Figure 3–19). As a giant iceberg melts, it continues to bob up above sea level, keeping pace with its melting. But neither mountain nor iceberg attain their initial elevation, for the volume continually diminishes.

Isostatic uplift of eroding mountains continually exposes more and more of the previously submerged mountain core. This process accounts for the present-day outcroppings of rocks once deeply buried tens of kilometers in the earth's crust. Isostatic adjustment during erosion also explains the enduring topographic relief of mountains. Erosion rates are rapid enough that even the highest mountains would be leveled in a few tens of millions of years without the compensating uplift.

Maintenance of isostatic balance is evident in other cases. During the glacial ages, large portions of the northern hemisphere covered by ice sheets more than 1000 meters thick were isostatically depressed by the great weight of these ice masses. About 12,000 years ago, the ice sheets melted and allowed the depressed crust to rebound isostatically, rising upward a few hundred meters. Continental margins also responded isostatically: melting glaciers flooded the edges of continents, sinking the inundated areas a few tens of meters. During glaciation, on the other hand, water volume in the seas decreased as ice built up, and the continental margins were drained of some of their ocean water. The edges of the continents then rose isostatically as the weight of the overlying water was reduced.

On a much smaller scale, earth scientists have recorded changes in ground elevation behind dams. For example, the huge mass of water stored in Lake Mead, behind Nevada's Hoover Dam, has caused the land to subside by several centimeters. Once the isostatic adjustment is completed, total subsidence will be several meters. The opposite happened with Lake Bonneville, which covered most of Utah's western half during the glacial age. When glacial precipitation diminished and the lake began to evaporate, it shrank to one-twentieth its former area; now we call it Great Salt Lake. After all this water disappeared, the center of the area formerly covered by Lake Bonneville rose several tens of meters. (The salty character of Great Salt Lake is due, in part, to the evaporation of Lake Bonneville, whose dissolved salts were left behind. Salts continue to concentrate today because Great Salt Lake has no outlet, and all dissolved substances carried there by streams accumulate there; most of the water evaporates in the arid climate.)

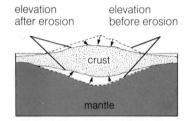

Figure 3–19

Highly simplified sketch of isostatic uplift of continental mountains accompanying their erosion. Dashed lines indicate continental mass before erosion; arrows show uplift of lower and upper continental surfaces during and after erosion. Elevation of continental mountains following uplift is less than original elevation because erosion has reduced the total volume of the continental mass.

Isostatic adjustment proceeds at a snail's pace. Loading (by sediment) and unloading (by erosion) of the earth's crust take place over thousands to millions of years. Restoration of isostatic balance is equally slow because of the viscous nature of the underlying mantle. Isostasy, then, like plate tectonics, keeps the earth's crust in constant though creeping motion.

3–6 Another Look at Crustal Structures

Now we can survey the major geologic structures within the earth's crust, for these are fashioned by the processes we have just described. It should be apparent that the array of continents, ocean basins, and zones of volcanic and earthquake activity is not a random one. Rather, their arrangement relates closely to the presence and movement of the lithosphere's rigid plates as portrayed in Figures 3–20 and 3–21.

Figure 3–20

Major crustal structures of the earth and their relation to earth plate boundaries.

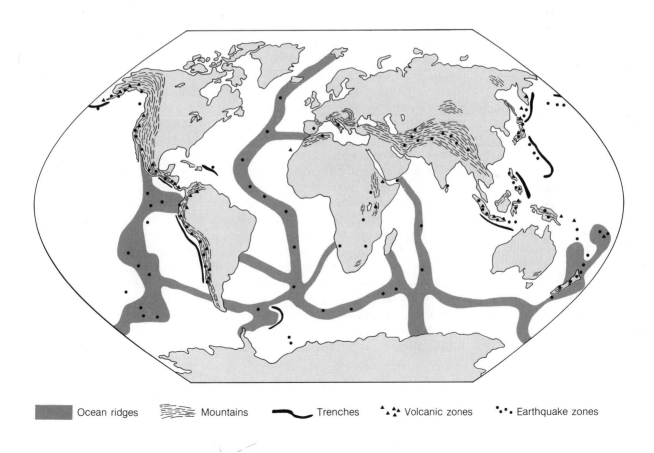

Ocean ridges   Mountains   Trenches   Volcanic zones   Earthquake zones

Figure 3–21

Idealized cross section of
boundary between oceanic and
continental plates. When plates
collide, less dense continental
plate rides over denser oceanic
plate that plunges down into
asthenosphere. Deep-sea trench
forms at boundary, and oceanic
sediments either pile up against
leading edge of continental plate
or ride down with oceanic plate.
Thick wedge of sediments eroded
from continent accumulates along
its trailing edge and creates wide
continental shelf and slope.

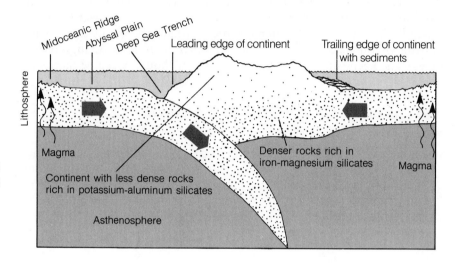

Continents and ocean basins

About one quarter of the earth's surface, the continents, lies above sea
level. Most of this area is underlaid with enormous masses of light rocks
rich in silicon and aluminum. Here the lithosphere is several tens of
kilometers thick and floats on the denser asthenosphere. The continents
are uplifted by tectonic and isostatic processes and are continuously eroded
by wind, water, and ice.

Each continent has one or more core areas, or *shields*, of ancient,
crystalline igneous and metamorphic rocks. Possibly the shields represent
proto-continental orogenic belts that formed early in earth history. They
are surrounded by younger, deformed, and metamorphosed orogenic belts
that create long, narrow mountain systems.

The continents thin and taper toward the oceans. On the rim of the
Pacific, continental margins drop abruptly into the ocean basin; on the
rim of the Atlantic and most of the Indian Ocean they slope gradually
below the sea. Thus Pacific continental margins are narrow, falling quickly
into adjacent deep sea trenches. Atlantic and most Indian Ocean continental margins are broader and not bordered by trenches; instead, they
have a thick wedge of sediments called the *continental rise* that has been
eroded from the adjacent landmasses.

Approximately three-quarters of the earth's surface lies below the sea.
Ocean basin rocks are denser and considerably thinner than the rocks
of the continents; hence the basins sit lower on the underlying mantle.
Because they are topographically lower than the continents, the ocean
basins receive most sediments eroded from the land.

Long basaltic ridges in the central portions of ocean basins mark the diverging plate boundaries where basaltic magmas accumulate—take another look at Figure 3–20. Away from the ridges are the broad, deep-water, submarine *abyssal plains* where much of the ocean's sediment slowly settles. At some converging plate boundaries, on the other hand, sediments may be scraped off the ocean floors and piled against the continent or carried down along the subduction zone with the descending plate. Deep sea trenches at the leading edges of continental plates mark the location of these downward-moving plates. Continental rises, the thick wedges of sediment eroded from the continent, accumulate on the trailing edge of the continental plate.

Volcanic and earthquake zones

As noted earlier, a huge network of active volcanoes traces the plate boundaries. In general, the volcanoes at diverging plate boundaries erupt basaltic lavas from the mantle below, which is rich in iron and magnesium silicate. At converging plate boundaries, the volcanoes produce lavas that are more rhyolitic. This is because the latter magmas come from sediments and rocks on the descending plate in the subduction zone which are rich in potassium-aluminum silicates.

Earthquakes generated by plate movement are also associated with plate boundaries and the volcanic activity occurring there. Shallow-focus earthquakes are characteristic of midoceanic ridges, whereas deep-focus earthquakes also occur along subduction zones where an ocean plate moves under the edge of a continental plate.

Viewpoint   Peter A. Rona

Peter A. Rona is a research scientist with the National Oceanic and Atmospheric Administration, Miami, Florida. Dr. Rona's recent work has been on the possible link between plate tectonics and economically valuable mineral deposits. In this Viewpoint Dr. Rona reviews the geologic factors controlling such deposits along plate boundaries.

Learning Where to Look for Mineral Resources

The revolutionary advance in our understanding of internal processes of the earth made during the past ten years is leading to economic returns through the discovery of new mineral resources. The problem of exploring the earth's crust for mineral deposits is the proverbial one of searching for a needle in a haystack. Just as the needle is tiny relative to the size of the haystack, so petroleum and metal deposits are small relative to the volume of the earth's crust. Many valuable mineral deposits occupy areas less than that of a city block. The problem of discovering a deposit, like that of finding the needle in the haystack, is facilitated by knowing where to look. In the past we have found only the most accessible deposits on the continents, largely by trial and error, without a real understanding of why and where the deposits occurred. As a result of our increased understanding of internal processes of the earth we are gaining insights that tell us where to probe for new sources.

Seem like a good place to look?

How do internal processes control mineral deposits? An important clue is the observation that many minerals lie along the boundaries between the plates that divide the earth's crust (Figure 3–1). Petroleum and various base and noble metals occur along the margins of the Pacific Ocean, which are plate boundaries. These deposits include the metal provinces of western North and South America, and the petroleum of Indonesia and the west coast of the United States. We are exploring the plate boundaries submerged beneath the ocean along midoceanic ridges and have found that overlying sediments are enriched in various metals and that solid metal deposits appear at certain sites along the Mid-Atlantic Ridge. For this reason we suspect that metal deposits including copper, manganese, iron, nickel, lead, zinc, chromium, cobalt, uranium, and gold may be present at sites on midoceanic ridges. We do not think that midoceanic ridges contain any petroleum. Sediments on the floor of the Red Sea, where a plate boundary lies between Africa and Eurasia, are enriched in iron, zinc, copper, lead, silver, and gold. Sediments full of organic matter that may eventually become petroleum are preserved at the margins of the Red Sea.

Why should certain mineral deposits be concentrated along plate boundaries? Plate boundaries are where the geologic action takes place. As crustal material is being created at divergent plate boundaries and is being destroyed at convergent plate boundaries, processes are working to concentrate minerals in deposits along the boundaries. The metals that appear at divergent plate boundaries on midoceanic ridges and in the Red Sea are being deposited from hot solutions that concentrate metals from the rocks lying at these boundaries (Figure 3–2a). The metals at convergent plate boundaries around the Pacific Ocean are also deposited from hot solutions rich in metals possibly derived from the melting of the Pacific plate as it plunges beneath the adjacent continents (Figure 3–2b).

Many mineral deposits lie in areas far from present plate boundaries. To understand the reason for the location of these deposits, we must consider how the sizes, shapes, and positions of the continents and ocean basins have changed through time. For example, the Atlantic originated as a sea at an early stage in the development of the divergent plate boundary and widened into an ocean over a period of 200 million years. Various metals and organic matter may have accumulated in the Atlantic Sea, as they have in the present Red Sea, so that metal and petroleum deposits may be present at sites under the miles-thick sediments along the eastern and western margins of the Atlantic Ocean.

The separation of continents by continental drift about divergent plate boundaries may divide preexisting mineral provinces. Fitting the continents together in their positions prior to continental drift in a global jigsaw puzzle may reveal the continuity of certain mineral provinces between the continents. Diamonds found in northeast South America (Guyana and Venezuela) appear to have been derived from source rocks in west Africa (Liberia, Ivory Coast, and Ghana) when South America and Africa were joined prior to the opening of the Atlantic Ocean. Gold-bearing formations in these two regions of South America and Africa can be matched across the Atlantic. Metal provinces of southeast Africa, southern India, and western Australia apparently match as well, agreeing with the positions of these continents when joined prior to continental drift.

Mineral deposits may also be present along former plate boundaries that are no longer active. Metal deposits of the Appalachian Mountains of southeastern North America and the Ural Mountains between Europe and Asia may have originated when these mountain ranges developed at former convergent plate boundaries. Still other mineral deposits within continents do not appear to conform either to present or former plate boundaries, and their origins remain problematic.

The patterns of mineral distribution that are emerging from our increased understanding of internal processes of the earth are guiding man's search for new mineral deposits. However, many factors should be kept in mind concerning mineral resources for the future. Nature is not manufacturing new mineral resources at plate boundaries as fast as man is depleting known mineral deposits. The rate of accumulation of metal deposits from hot solutions at midoceanic ridges is about 8 millionths of an inch per year (2 hundred-thousandths of a centimeter per year). The minimum time required to form petroleum from organic matter and concentrate it into pools is of the order of tens of thousands of years. Many existing mineral deposits are inaccessible to us because they are too deeply buried in the crust to be detected and recovered by present technology; and when deposits are accessible, often the cost of production would be higher than the market value. Even with our increased understanding of where to look for new mineral deposits, evaluation of a new area of land or sea may require years of exploration and expenditure with uncertain outcome. Following the discovery of a valuable petroleum or metal deposit, five to ten years of developmental work are generally required before the prospect is ready for production. In conclusion, our greater understanding of the earth's internal processes can be expected to accelerate the discovery of mineral resources on the continents and in the ocean basins, but still offers us no promise of utopia.

Summary

The earth's present-day surface results from large-scale processes operating throughout its crust and upper mantle. The all-encompassing theory of plate tectonics explains the broad features and geologic activity that we observe on the face of the earth. Relatively thin, rigid plates move either by gravity or convection over the thicker, plastic upper mantle. Plate boundaries are marked by long, linear geologic features like mountain ranges, submarine basaltic ridges, volcano and earthquake zones, and deep sea trenches—depending on whether plates are moving together, moving away from each other, or slipping past one another.

Continents are thick slabs of lighter rocks floating more or less in isostatic balance on the denser, underlying mantle. Ocean basins are underlaid with thinner slabs of heavier rock resting on the mantle. Continents are regions of higher elevation continually eroding; ocean basins are areas of lower elevation that receive sediments deposited from the continents. As oceanic plates collide with continental plates, the oceanic sediments may be scraped off and piled against the continents or metamorphosed and granitized as they descend below the continent. Plate collisions also deform rocks by faulting and folding them.

Basaltic lavas continuously well up from the mantle and form long, linear ridges within the central portions of ocean basins along diverging plate boundaries. Plate movement away from the midoceanic ridges brings these mantle materials from the oceans and eventually adds them to the continents. Plates thus grow at their diverging boundaries and are consumed at converging ones. Consequently, throughout geologic history continents have grown with the continuous addition of mantle material and have shifted spatially as plate movement has carried them along.

Glossary

asthenosphere   Weak layer within the mantle lying below the lithosphere that behaves as a viscous liquid.

epicenter   Point on the earth's surface directly above the focus of an earthquake.

fault   Plane of breakage within the earth along which rocks have moved differentially.

focus   Point of initial rupture within the earth that results in an earthquake.

isostasy   State of balance in the earth's crust and mantle whereby rocks achieve different elevations according to their relative densities.

lithosphere   Strong layer within the crust and upper mantle above the asthenosphere that behaves as a rigid solid.

orogeny   Formation of mountain systems, particularly by folding, faulting, and igneous activity.

plates   Individual portions of the lithosphere that diverge, converge, or shear past one another as they ride along the underlying asthenosphere.

remanent magnetism   Magnetization of rocks induced by earth's magnetic field during their formation or subsequent heating.

subduction zone   Region where one plate pushes down below another plate as the plates converge.

tectonics   Structural behavior of the earth's crust and mantle including plate movements, earthquakes, orogeny, and crustal uplift and subsidence.

Reading Further

Clark, S.P., Jr. 1971. *Structure of the Earth*. Englewood Cliffs, N.J.: Prentice-Hall. A short, concise discussion of the earth's magnetism, plate tectonics, earthquake activity, internal physical and chemical composition.

Gordon, R. B. 1972. *Physics of the Earth*. New York: Holt, Rinehart and Winston. Companion volume to *Chemistry of the Earth* (cited in Chapter 2) that discusses physical processes operating within and on the earth.

Wilson, J. Tuzo, ed. 1972. *Continents Adrift*. San Francisco: W.H. Freeman. A collection of articles describing the processes and results of plate tectonics. Most are written by the people who made the original discoveries.

Surface Processes 4

So far we have looked at our planet from a large-scale, global perspective. We have examined its initial formation and subsequent differentiation, its minerals and rocks, and the internal processes that keep it in constant flux. In this chapter we discuss the surface processes that shape the earth's landscapes. During this shaping, the earth's solid outer shell interacts with its fluid shells of air and water. Wind, water, and ice chemically alter and physically disintegrate rocks, move them downhill, and deposit them at lower elevations where eventually they solidify into sedimentary rocks.

The energy that drives surface processes is of two kinds: solar and gravitational. Solar energy activates the hydrologic cycle whereby water is evaporated from land and sea, falls as rain and snow, and returns to the sea. Gravitational energy is the force that moves water and ice downhill, carrying with them rock debris eroded from the earth's surface.

All the rivers run into the sea; yet the sea is not full.

Ecclesiastes, c. 280 B.C.

4-1 Weathering

Rocks and minerals in the solid crustal shell interact with the earth's biosphere as well as with the fluid shells of air and water. The first step in this interaction is *weathering*, the response of earth materials formerly in equilibrium with chemical and physical conditions within the crust to new conditions on the earth's surface. These new conditions include lower temperatures and pressures; abundant water, oxygen, and carbon dioxide; and animal and plant activity. Also, these conditions usually vary widely from day to day, seasonally, or over the years.

A change of chemistry

Water from rain, snow, and dew mixes with oxygen and carbon dioxide gas in the atmosphere, as well as with various acids in soils. After the water is mixed, it becomes a chemical solution that decomposes rocks and alters their minerals through dissolution, oxidation, and hydration. See Figure 4–1. *Dissolution* occurs when carbon dioxide from the atmosphere or soils mixes with water and forms a weak acid, carbonic acid (H_2CO_3). The slightly acidic surface water dissolves some elements within minerals and carries them off in solution to underground water, rivers, and oceans. *Oxidation* happens when oxygen in surface water combines with elements in minerals to make new oxide compounds. *Hydration* involves the chemical combination of water itself with other elements. For example, the chemical interaction of orthoclase feldspar with surface water eventually produces clays and leaves silica, potassium, and bicarbonate ions in solution. Chemical weathering of biotite mica yields the new minerals of goethite and chlorite, and also places silica, potassium, and bicarbonate ions in solution.

In short, chemical alteration creates new minerals from old, and releases numerous dissolved chemical elements into surface water. The dissolved substances are carried by streams, rivers, and underground water to ponds and lakes, and eventually to the ocean itself. On their way to the ocean, some dissolved substances may be precipitated from the solution when they pass through rocks and sediments. Upon reaching the sea, some may be precipitated as new minerals by inorganic processes or by organisms making their shells. Finally, the dissolved substances may precipitate within clastic sediments, cementing them into hard rocks. Alternately, removal of these substances from seawater by evaporation or by animals and plants is responsible for nonclastic sediments (refer to Section 2–3).

The susceptibility of minerals to chemical alteration depends on their order of crystallization from a magma, which we dealt with in Section

Figure 4–1

Chemical weathering of limestone in Maryland. Removal of more soluble limestone and oxidation of iron-bearing minerals leaves behind dark, insoluble material and clays originally incorporated within the rock.

2–2. Minerals like olivine, pyroxene, amphibole, and plagioclase feldspars, which have crystallized early, are less resistant to chemical weathering than the later-crystallizing minerals like micas and orthoclase feldspars. In general, minerals with higher silica content are more resistant; quartz, which is composed entirely of silica, is the most resistant of all.

Little rocks from big rocks

Rocks and minerals disintegrate physically when they mechanically break down to smaller sizes and there is little or no chemical alteration. The individual particles or mineral grains that physically compose rocks separate from one another and cause the rocks to break apart. Igneous and metamorphic rocks are bound together into a rigid, dense mass by their interlocked mineral crystals. Sedimentary rocks are also lithified by the crystalline cement that binds chemically precipitated, nonclastic or clastic grains. Usually physical weathering fragments these rocks along crystal or particle boundaries and makes little rocks out of big ones.

When rocks crystallize, they usually do so under considerable pressure, whether as granites in the core of a mountain range, as shales beneath a pile of thick sediments, or as schists buried within the crust. As these rocks are brought to the surface, where pressures are less than those under which they initially crystallized, the rocks expand and fracture from the release of stress. For example, in large granite masses like batholiths, extensive fractures develop parallel to the ground surface as erosion slowly unroofs the batholith's overlying rocks (Figure 4–2). Such fracturing is called *exfoliation*—the fractures result in a peeling away of leaves or layers from the outer surface of the massive rock.

Similar fracturing of massive rocks occurs in quarries; as rock is removed the remaining rocks may spontaneously fracture when they expand.

Figure 4–2

Exfoliation produced in granite in high Sierras of California from release of pressure as overlying rocks have eroded. Exfoliation is parallel to the ground surface; scattered trees have colonized in some of the crevices of the granite, and their growing roots will contribute to further fracturing of the rock.

After rocks are fractured, water *percolates,* or seeps down, into them and may freeze. Within rock crevices, the expansion of water when it freezes to ice further fragments the rocks and thereby intensifies physical disintegration. (The water may also, of course, contribute to their chemical alteration.) Unusual fluctuations in temperature from desert sun, forest fires, or arctic cold may also cause enough rock expansion or contraction to fracture and break them into smaller pieces.

Weathering action by living things

Animals and plants live on rocks of all sorts and contribute to their weathering. On land, plants as diverse as lichens, grasses, and pines grow directly on rocks or in their crevices. Plants may root themselves in newly formed fissures, and their growing roots then add to the rocks' continuing fragmentation. When plants decompose, organic acids mix with surface water and aid in chemically altering the rocks.

Along coastlines, marine organisms like blue-green algae, clams, chitons, sea urchins, and barnacles bore into solid rock. These borings weaken the rocks and make them more susceptible to other kinds of erosion, such as wave attack from the surf. Other animals, especially snails, graze on the algae that coat coastal rocks and chew off a thin layer of rock.

Weathering processes are closely tied to the prevailing climate. The rate and nature of weathering differs, for instance, between the warm, humid

Figure 4–3

(a) Swiss Alps, where repeated freezing and thawing facilitate active physical weathering. Except for some mosses and lichens, plants are absent due to extreme temperatures and thin soils. (b) Tropical forest in warm, humid climate; chemical weathering is very active in such an environment. Lush vegetation is supported by equable climate and thick soils.

(a)

(b)

climate of a tropical forest and the cold, dry conditions of a high-latitude desert. In general, warm temperatures and plentiful water favor chemical and biological weathering, whereas cold temperatures and scarce water favor physical weathering (Figure 4–3).

The chemical, physical, and biological disintegration of rocks and minerals produces loose surface deposits, or *soils*, in which plants and animals live. The character of the soil reflects both the composition of the underlying rock and the dominant weathering process. Further weathering ensues when microorganisms like bacteria and fungi colonize the soil and secrete organic acids that slowly attack the rock below.

4–2 Gravity's Pull

When rocks and soil move downslope—pulled by the force of gravity and only a little help from wind, water, or ice—we have another shaping process: *mass wasting*. Contributors to the process include such phenomena as soil creep, solifluction, slumps, slides, mudflows, and debris avalanches. But before discussing all these types of mass wasting, we should take a look at the role of gravity.

The attractional force between the earth's center of mass and objects on its surface generates *gravitational energy*. This force is constant and continuous, and its net effect is to move objects from higher to lower elevations—that is, from positions farther from the earth's center of mass to positions closer to it. Some objects can resist this movement because of their internal strength or because of the frictional force between the object and ground surface on which the object rests. Mount Everest, for instance, does not collapse under the force of gravity because the internal strength of its rocks holds it together. Nor does my car roll downhill when the parking brake is on, for the frictional forces between the tires and the road are greater than the gravitational force pulling the car downhill.

We intuitively realize that the steeper the slope of the ground, the greater the tendency for an object to roll or move down that slope. If you park on a level spot, the car won't roll if the brakes are off; but when you park on a steep hill, you know instinctively to put the car in gear, turn the wheels into the curb, and pull up tightly on the parking brake. The downslope energy which drives mass wasting is gravitational in origin and thus proportional to the slope on which surface materials lie, as outlined in Figure 4–4.

Soil creep and solifluction

Slow downslope movement of relatively dry soil and loose rock debris is called *soil creep*. The downslope component of gravity always exerts

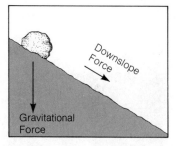

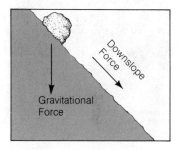

Figure 4–4

Diagram illustrating the relation between the force tending to move an object downhill and the slope on which it lies. The tendency for objects to move downslope increases with slope. The frictional force between object and slope resists the downhill force. Stable rocks on a dry slope, for example, may slide down that slope when wet, because the water reduces the friction-resistance.

a force on such material, but frictional forces inhibit movement. Yet any time soil and rocks are disturbed, some movement occurs, no matter how slight. As an example, soil and rock debris heave outward, perpendicular to the slope, when frozen. And when they thaw, they do not return to precisely the same place on the slope, but are displaced by gravity a small distance downslope. In a similar way, when rain or hail strike a slope, when hikers or cattle tramp by, when clays swell from wetting and shrink with drying, when a tree falls, or when an animal burrows, all these movements dislodge materials in the loose ground surface a little bit further downslope. So the loose, granular soil and rock fragments that carpet hillsides move slowly and inexorably downslope, and the steeper the slope, the faster the creep.

Solifluction is similar to creep, except that soil and rock debris are saturated with water and flow more as a unified mass. In areas underlaid with an impervious layer of material, such as a shale bed or perenially frozen ground, called permafrost, solifluction is common because surface water cannot penetrate very far and instead accumulates in the loose surface materials.

The effects of creep and solifluction are evidenced by tilted fence posts and telephone poles, and by cracks in roads whose shoulder is moving downslope. As in Figure 4–5, trees on hillsides are often curved at their bases because of creep; as they inch downhill, their trunks grow upward and backward toward the sunlight overhead.

Slumps and slides

When discrete, isolated masses of soil and rocks move down a well-defined surface of slippage, we have *slumps* and *slides* like those in Figure 4–6. Slumps rotate backward toward the slope with only slight downslope movement of material. A slide detaches itself completely and moves further downslope, often breaking apart and disintegrating as it goes. Slumps and slides move considerably faster than do creep and solifluction, as well as differing in their complete detachment from the underlying surface. The size of slumps and slides ranges from small clumps a meter or so across, to huge displacements thousands of meters long. Slumps and slides often happen after heavy rains or melting of thick snows: the soil and rock mantle's internal strength is reduced by water, or the plane of potential slipping is lubricated by the water.

As you can see in Figure 4–6, the appearance of landslides is readily apparent, so scrutiny of a landscape may reveal a past history of sliding and slumping. Some areas may not normally be prone to sliding or slumping until some unusual event provides the appropriate conditions. Forest fires, for example, may remove the ground vegetation that ordinarily inhibits such mass wasting. Later, heavy rains may so saturate the upper ground surface that extensive slumping and sliding begin. The structure of the

Figure 4–5

Trees moving downslope with soil creep in Sierras of California. Bases of trees are curved because trees grow upward toward light, compensating for downslope tilt due to creep.

(a)

Figure 4-6

Slump and slide in Montana. (a) Head of large slump on mountain slope where soil and rock debris have dropped about 10 meters vertically along slippage plane. (b) Massive landslide triggered by earthquake rushed down slope on right and continued up opposite slope some 100 meters. Note buried highway and dry creek bed; the slide impounded waters forming a lake more than 6 kilometers long behind it.

(b)

underlying rock, too, may facilitate downslope movement. If joints and rock strata are parallel to the slope, slides and slumps can more easily detach themselves from the bedrock and move downslope along the joint surface or bedding plane.

Rockfalls and mudflows

Two additional types of mass wasting occur. Although both move fairly rapidly, they differ greatly in their water content. *Rockfalls* are chunks

How many examples of mass wasting can you identify around where you live?

of rock that suddenly break away from a steep slope or vertical cliff face, hurtling through the air and landing at the slope's base. *Mudflows* are viscous masses of soil and rock debris moving rapidly down a slope. A mudflow is more watery and faster-moving than solifluction. Its path is also more confined, usually following a preexisting stream channel. Sudden torrential downpours in desert regions, for instance, may flow quickly down a dry canyon, picking up loose ground materials as they move along and making a slurry of mud and rock debris. Water seepage into the dry stream bed makes the flow increasingly stickier and more viscous as it continues down the canyon (Figure 4–7). Figure 4–8 summarizes the common types of mass wasting phenomena in terms of speed and water content.

Figure 4–7

Mudflow in a creek bed that ended up in these people's backyard in southern California. A sudden downpour in this semi-arid environment swept up loose soil and rock debris, forming a mudflow. Loss of water by seepage into the dry creek bed made the mudflow increasingly viscous until it became totally immobilized, but not before it destroyed many houses built in the area.

4-3 Stream and River Transport

The most significant and obvious surficial process is the runoff of surface water across the land accompanied by erosion and transportation of sediments. When rain falls, vegetation, soil, and rocks on the ground surface absorb some of the moisture. Depending on several factors, including soil depth, nature of underlying rock, and rate of precipitation, the ground becomes saturated and the water begins to flow overland, moving toward lower and lower elevations. Melting water from ice and snow, of course, moves downslope, too. This slope water initially moves in a thin continuous

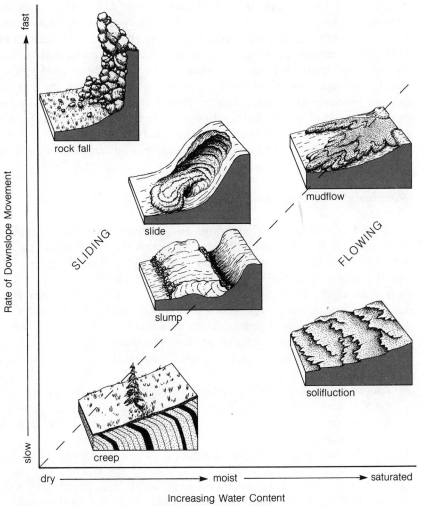

Figure 4–8

Common types of mass wasting defined by rate of movement and water content. Creep, solifluction, and mudflow travel by flowing while maintaining contact with the underlying surface; slumps, slides, and rock falls move by sliding and breaking contact with the underlying ground.

sheet, and soon begins to erode the ground surface. In doing so, it forms small linear depressions, or *rills*, that after a short distance connect and form streams and eventually rivers. The land is thus covered with a branching network of channels of all dimensions. The network begins with numerous rills at higher elevations that become rivers that empty into lakes or the sea. Throughout its journey, this surface water carries material eroded from the land—dissolved substances, suspended particles, and rolling grains—much of which is deposited at the river's mouth.

The geographic area where precipitation accumulates is called the *watershed;* its outer periphery stretches along the crests of hills, ridges, and mountains down which the water flows. Below the higher elevations that define the watershed lies the *drainage basin*, the area draining the watershed through its network of streams and rivers. (The watershed of the Mississippi River system includes part or all of twenty-eight states!) Not all water within the drainage basin flows over the ground surface. Some water percolates down through the soil and underlying rocks, replenishing the supply of water that lies at varying depths below the ground surface. Like the surface flow, groundwater moves downhill through the soil and through porous rock strata called *aquifers*. Groundwater may again find its way to the land surface farther downslope in the form of springs feeding a brook, pond, stream, or lake. Near the coasts, groundwater may reach the sea directly by flowing into the marine groundwater that penetrates the subsurface along the coast.

Watching the river run

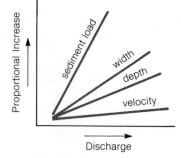

Figure 4–9

Proportional changes in sediment load, width, depth, and velocity in streams and rivers with increases in discharge. Data from which the graph was drawn come from several river systems in the United States representing a variety of geographic locations, relief, local geology, and size.

Most land is sculptured when water and sediment flow in stream and river channels. Before the sculpture can proceed, though, the land surface must be prepared by weathering and mass wasting. The channels along which water and sediment travel can be individually characterized by their *discharge* (i.e., the amount of water flowing past a certain point during a given period), their *width* and *depth*, their *velocity*, and their *sediment load*. Studies of numerous streams and rivers have yielded some important relationships among all of these characteristics:

- Width, depth, velocity, and sediment load increase with an increase in discharge.
- Rate of increase of width, depth, and velocity is somewhat *less* than the rate of increase of discharge.
- Rate of increase in sediment load is considerably *greater* than the rate of increase in discharge.

These relationships are detailed in Figure 4–9. In a qualitative sense, they indicate that during heavy surface runoff when the discharge rises, the channel will correspondingly increase in width and depth, the water velocity will accelerate, and the sediment load will grow. If the discharge is sufficiently large, the channel will erode and so become still wider and deeper. If channel erosion does not accommodate the discharge, the stream or river will overflow its channel banks and flood its valley.

Another way to visualize these relationships is to consider the changes occurring downstream within a river system during normal flow. Upstream, the system contains many small tributaries. As these individual streams connect into larger streams, the discharge increases, as do channel width, depth, velocity, and sediment load. Continuing farther downstream, larger tributaries eventually coalesce into the main river; its channel holds the aggregate flow of all smaller, upland streams and its discharge is greatest. For this reason, the river's channel width, depth, velocity, and sediment load are proportionally greater.

Sediment load

As water flows over the ground surface, it carries soil particles and rock fragments with it. Mass wasting upslope of the stream and river margins also contributes sediment to the channels. The sediment load in streams and rivers is of three kinds: dissolved, suspended, and tractional. The *dissolved load* includes all the substances in solution in stream and river water. Many of these substances have been dissolved from minerals and rocks by chemical weathering; others come from organic materials produced by animals and plants. There are also dissolved wastes intentionally or accidentally dumped by humans.

The *suspended load* includes fine-grained sediments, mostly clay and silt. Because of their small mass and large surface area, the sediments are easily kept in suspension by the turbulence of the flowing water. The *tractional load*, sometimes called the *bed load*, consists of sediments pulled along by frictional dragging, or traction, of the flowing water in contact with the stream or river bed. Particles in the bed load are sand, pebbles, and cobbles, and they are rolled and bounced downstream by the current.

A stream or river's *competence* is the maximum size of particle it can carry; the *capacity* of a stream or river is the total amount of sediment load. Competency rises with velocity—a flooding torrent can carry rocks and felled trees that a placid river could not. Capacity grows with total discharge as well as velocity—the lower stretches of the Mississippi carry far more total sediment than any of its tributaries in Wyoming or Montana. The relative proportions of dissolved, suspended, and bed load vary from stream to stream. In general, most of the sediment load is suspended, accounting for half to more than three-quarters of the total.

Why do you suppose that this is so?

4-4   Streams and Rivers at Work

Flowing water lowers the land surface by eroding and transporting soil and rock, complementing and completing the work done by weathering and mass wasting. Sediments carried downslope by mass wasting and the sheet flow of water form a veneer of loose materials that is generally continuous over the slopes. Farther downhill, such loose materials are accumulated by flowing water and contained in larger volumes in stream and river channels. This flow of water shapes landscapes in different ways, depending on the relative supply of water and sediment, rates of weathering and mass wasting, and character of the materials being eroded.

Alluvial fans and braided rivers

In regions where the sediment supply is greater than the water flow, the excess sediments are deposited along the stream and river channels, building them upward and raising their elevation. This adjustment is called *aggradation*. For example, in mountainous areas with arid or semi-arid climates, where physical weathering produces large amounts of coarse sediment, the meager precipitation cannot transport all the surface debris. In such areas the local streams and rivers create two kinds of aggraded landforms: *alluvial fans* and *braided rivers*.

Let's first consider alluvial fans. As you can see in Figure 4–10, alluvial fans are cone-shaped sediment accumulations lying on valley floors at the foot of mountain slopes. Intermittent streams, flowing only during torrential rains or the rainy season, transport much of the available sediment

Figure 4–10

Vertical aerial view of coalescing alluvial fans, Death Valley, California. Notice the cone-shaped outline of the fans with their apexes pointing into canyons running out from the mountains onto the valley floor; note, too, the numerous stream channels that take turns building up the fan.

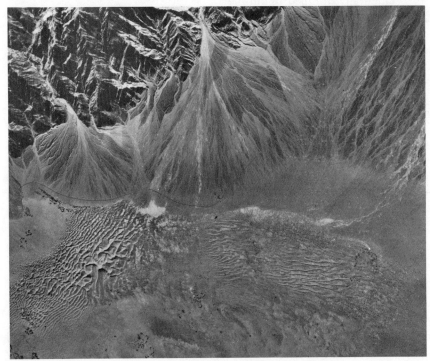

down steep canyons cutting across mountain slopes. When the streams reach the neighboring valley floor, the water velocity drops and the stream's competence and capacity sharply decline. The sediment carried by the stream is rapidly deposited as a layer of debris at the canyon mouth. Over time, these deposits build up as an alluvial fan, a pile of sediment whose apex points toward the canyon. Finer-grained materials may remain in suspension and be deposited on the valley floor farther out from the canyon. Occasionally a temporary lake will form, only to evaporate when the dry season comes.

Alluvial fans develop only in arid and semi-arid regions, but braided rivers form in other climates as well, just so long as the seasonal or longer-term discharge fluctuates enough. In the wet part of the year, or during a rare interval of heavy rainfall, large quantities of sediment gathered from weathering and mass wasting will be carried along a stretch of the river channel. But when the rain lets up, such heavy discharge can not be maintained, so much of the sediment brought into the channel is stranded there as the discharge wanes. The stranded sediment forms long, linear, interwoven sand bars among which the river subdivides as it flows along. Such rivers have been dubbed braided rivers because of their interlacing sandbars and multiple channels; refer to Figure 4–11.

Sediment stranded one time will be carried farther downstream at the next high flow, and its place taken by newly introduced sediment. Thus,

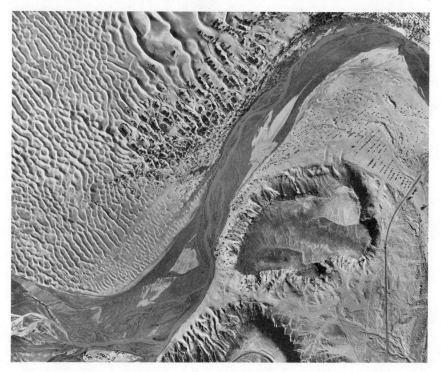

Figure 4–11

Vertical aerial view of braided river in Algeria. Notice the many sinuous sand bars and channels within the main channel of the river. (Can you identify other surface features?)

the channels in braided streams constantly shift from flood to flood, or rainy season to rainy season. In terms of their width and depth, braided streams and rivers are broad and shallow. The ratio of depth to width is therefore low, and reflects the need for greater traction along the channel floor to pull along coarse sediment composing the bed load.

Meandering rivers and floodplains

In regions where water flow is more than sufficient to transport the supply of sediment being produced, streams and rivers cut down and erode their channels. Such *degradation* occurs because the stream or river's capacity is greater than the load supplied by its tributaries. Where the climate is humid and where topographic relief is medium to low, streams and rivers flow throughout the year along sinuous courses. That is, where water is plentiful and where elevation between the channel's highest and lowest point does not vary much—usually less than a few hundred meters—sinuous *meandering rivers* form. As they meander, they pick up sediments and carry them to their mouths.

Important agents of degradation, meandering streams and rivers slowly shift their courses along their route, and erode a wide valley along their overland path. This is evident in Figure 4–12. These streams and rivers

Figure 4–12

Valley in Nevada cut by meandering Humboldt River. The river winds back and forth across the valley, eroding a wide path over time. The valley is temporarily filled with sediment deposited by the river during times of floods.

have characteristic channel forms and channel deposits. As the loose, valley sediment is eroded, channel banks are undercut and cave in. The river flowing around these cave-ins is redirected toward the opposite bank, where it again begins to erode the bank. As the river strikes and erodes one side of its channel, it forms a *cut bank* which is rather steep; as the river swings out and away from the other side of its channel, it deposits some of its load and creates a *point bar* of sediment. With time, the river swings through numerous bends and as it continues to erode, meanders throughout its valley. *Oxbow lakes* are sometimes formed by meandering rivers when a loop in the river is cut off from the main channel. See the illustration in Figure 4–13.

During times of greatly increased discharge, meandering rivers may overflow their banks and deposit sediment throughout the valley floor. This is because the water velocity suddenly slows when the flooding water escapes its confining channel. The coarse sediment in this so-called *over-bank flow* is deposited as natural *levees* that are narrow, linear ridges paralleling the river's course and lying adjacent to the river bank. The fine-grained sediment is carried farther out on the valley floor where it settles as the flooding waters recede. Both kinds of sediments deposited by the river along its valley floor form the river's *floodplain*.

The valley floor of a meandering river, therefore, is both cut and filled

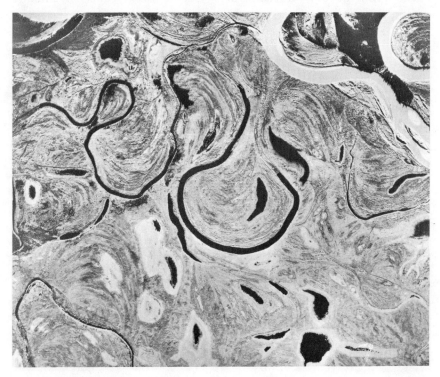

Figure 4–13

Vertical aerial view of meandering river in Alaska. Notice that some of the loops in the river channel have been cut off and isolated, forming oxbow lakes.

by the river itself. Sediments are only temporarily left behind, of course, because weathering, mass wasting, and surface erosion eventually convey the sediment down the valley. Anything left behind by these processes will be carried away by the river's future swings through the area. Meandering rivers, as well as those with less sinuous courses, have high depth-to-width ratios. Most of the sediment load is dissolved and suspended, and there is a much smaller bed load than in braided rivers.

Deltas

When rivers reach the sea, their velocity is checked, and they abruptly lose their competence and capacity. This leads to the deposition of the river's sediment load. The resulting sedimentary accumulation, a *delta*, is roughly cone-shaped with its apex pointing toward the river mouth, as in Figure 4–14. Sediment deposition works in a manner similar to the formation of an alluvial fan. In both, as the stream or river velocity rapidly drops, sediment is deposited. And, like an alluvial fan, as the river mouth shifts over time from side to side, it gradually builds up a cone-shaped accumulation. Furthermore, as with an alluvial fan, sediments are carried across the delta and down its front as it builds outward into the sea. So long as sediment continues to be brought to the delta, it will grow outward into the sea. If the sediment is for some reason reduced in amount,

If a river neither meanders nor is braided, what might you guess about its conditions of flow and the materials across which it moves?

Figure 4–14

Aerial view, looking south, of the Mississippi River delta at the edge of the Gulf of Mexico where one tributary of the river divides into three. Some of the sediment carried by the river is deposited within the channels (and adjacent to them during floods), but much of it is transported farther seaward where it settles out in the deeper waters of the Gulf.

seaward erosion by waves and currents will begin to destroy it. Deltas may also form on land where streams and rivers enter standing bodies of water like natural lakes or man-made reservoirs.

The graded stream

We have seen that streams and rivers modify their width, depth, velocity, and sediment load as their discharge changes. In aggrading, braided rivers, the build-up of the river bed makes its slope, or *gradient*, steeper and thereby speeds up its water velocity. The accelerated velocity, in turn, raises the river's competence and capacity to move the excess sediment downstream. Conversely, in degrading, meandering rivers, the downcutting of the river bed makes its gradient less steep; consequently, the velocity slows to a speed that is just sufficient to move the river's sediment.

In general, streams and rivers modify their channel geometry and gradient to adjust the water flow to their sediment load. Because a stream or river cannot control the amount of sediment or water brought to it, any imbalance is regulated within the channel itself. That is, the gradient changes to maintain a balance or equilibrium between water flow and sediment load. The concept of a graded stream or river—namely, the gradient's delicate and continuous adjustment to maintain equilibrium—is especially crucial to our understanding of how streams and rivers behave when human activities interfere with the natural processes influencing discharge and sediment load.

During construction, the sediment load of local streams usually grows larger. When vegetation is removed and the land is excavated for foundations, large amounts of surface material are exposed to erosion. Without any significant change in discharge, a local stream thus becomes choked with excess sediment. As such a stream builds up its channel to transport most of its extra load, the stream's character alters. What was perhaps a clear, through-flowing channel, is braided with excess sand and silt.

After construction, when excavations are filled and the area revegetated, sediment loads return to normal; yet water runoff and discharge may have been augmented significantly. Construction, of course, adds impervious land cover—roads, parking lots, driveways, roofs, patios, and so on—and thus even more runoff occurs. Water that would have previously percolated into the ground is now carried across these impervious surfaces, thereby increasing the discharge of the local stream. The stream is now *under capacity* (i.e., it has more discharge than load), and begins to erode its channel, which lowers the gradient and adjusts the increased discharge to the sediment load. Streams' attempts to adjust their gradients during and after construction are verified over and over again by the common observation of a natural stream course. First the stream becomes choked and braided with sediment during construction; then it erodes, carving wider and deeper channels, caving in its banks, and meandering.

The Viewpoint by Leopold at the end of the chapter discusses the impact of urbanization on stream quality and appearance.

4-5   Water at Work along the Coasts

Where land meets sea, waves and surf erode, transport, and deposit sediments. The sediments generated by this erosion, together with those brought to the sea by rivers, are distributed all along the coast. With time, seaward currents carry these sediments out into the ocean deeps, where they finally come to rest.

Waves and surf

When wind blows across the sea's surface, the water ripples and undulates. Strong winds blowing over a broad expanse of sea for long intervals of time create regular disturbances that rise and fall, forming successive crests and troughs. As you might guess, the geometry of these surface undulations is called *waves*. Although the wave shape moves downwind, the water particles within the wave travel in circular orbits and essentially in one place, with only a small net downwind movement of the water mass itself, as you can see in Figure 4–15. The diameter of these circular wave orbits decreases as the water becomes deeper. Hence a "storm-tossed sea" is something that happens mostly on the surface; water movement at a few tens of meters' depth is relatively trivial.

When waves approach shallow water, they begin to "feel bottom;" that is, the orbiting wave particles near the water surface begin to pick up sediments on the seafloor and move them in circular orbits. As waves advance shoreward, their orbits decay into ellipses and eventually the sediment is moved back and forth, along the seafloor. Except for waves generated by storms or earthquakes, most surface waves do not actually transport much sediment except in water shallower than about 10 meters.

As waves approach shore, they touch the seafloor and this creates friction on the bottom of the waves. Eventually the velocity differences between the upper and lower parts of the wave become so great that the wave's upper part overrides the lower part and spills forward in the graceful overreaching crest that we call *surf*. The energy expended by waves in shallow water on bottom sediments and by surf pounding the coastline both erodes and transports sediments and loose rock debris.

Cliffs and beaches

As waves roll in from the sea, their crests move in parallel lines. Approaching land, the waves feel bottom in shallow water, and frictional drag slows them down. But because the inshore, bottom topography is usually uneven, the parallel alignment of waves is disrupted and they begin to reflect sea floor contours. Along irregular coastlines, waves feel bottom farther offshore

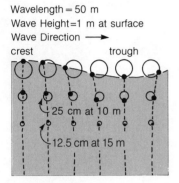

Wavelength = 50 m
Wave Height = 1 m at surface
Wave Direction ⟶

crest   trough

25 cm at 10 m

12.5 cm at 15 m

Figure 4–15

Geometry of waves. Water particles move in circular orbits that become increasingly smaller with water depth. Distance between crest and trough defines the wave height; distance between successive crests determines wavelength. Wave orbits are halved for every increase in depth equal to one-ninth the wavelength.

when they are opposite headlands and promontories than when they are opposite indentations, or embayments. This is because the higher elevations of headlands usually continue offshore, as illustrated in Figure 4–16.

(a)

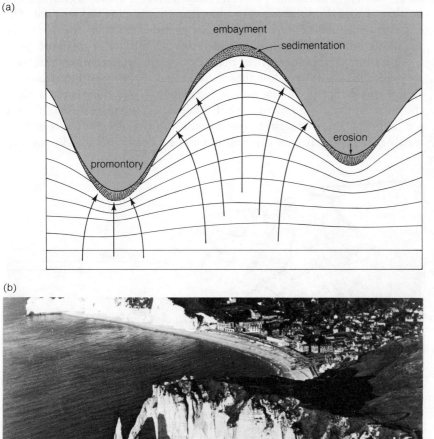

(b)

Figure 4–16

(a) Converging and diverging wave crests as they approach promontories and embayments. Waves opposite a promontory feel bottom farther offshore because the promontory's higher elevation continues offshore and makes the water shallower there. Opposite the embayment, water is usually deeper. The waves feeling bottom first are slowed more by friction while the waves in deeper water are slowed less. These differences in wave speed cause the waves to wrap around and converge on the headland; waves spread out and diverge in the embayment. The energy of wave and surf attack is thus concentrated along the promontories and dispersed along the embayments. Erosion is therefore dominant along promontories, and sedimentation dominant along embayments. (b) Rugged coastline of northern France with promontories interspersed with sandy beach embayments. Rocks are soft chalky limestones that erode relatively rapidly.

Waves approaching headlands converge toward each other, and their energy concentrates along a narrower front. On the other hand, waves approaching embayments diverge, and their energy is correspondingly spread over a longer stretch of coastline. Therefore, wave and surf attack is greater along headlands, and the subsequent erosion that creates cliffs is dominant. The relatively diminished wave and surf energy along embayments favors the accumulation of sediment there and forms beaches; refer to Figure 4–17.

Along more regular coastlines, the crests of approaching waves maintain their parallel alignment, but they may intersect the coastline obliquely. When this happens, it produces a *longshore current* that travels along the coast and takes sediment with it, as depicted in Figure 4–18. Offshore sediments, or those brought to the shoreline by streams and rivers or mass wasting, can be carried in this manner along a coast for many kilometers.

Figure 4–17

Schematic diagram of some important coastline features along steep and gentle coastlines.

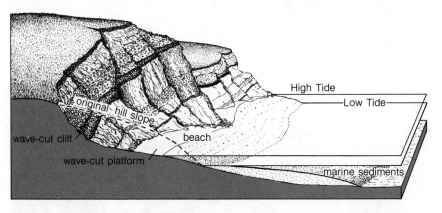

(a)　Steep Coast

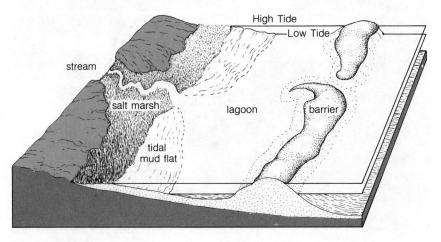

(b)　Gentle Coast

Figure 4–18

Wave crests from upper left strike Alaskan coast obliquely, generating longshore current parallel to shore that flows from top to bottom of photograph.

Although the beaches formed by longshore currents appear relatively permanent to a casual observer, their sediments are continually moving down the coast and being replaced by sediments from up the coast. Human interference with the natural sediment budget along such coasts can have startling and unexpected results. For instance, if we build jetties, dredge ship channels, or alter the sediment load and discharge of coastline streams and rivers, we can bring about rapid beach erosion or harbor silting far from the source of interference.

To what extent do these coastal processes also apply to the shores of lakes?

A simple example of coastline changes caused by man-made structures can be seen in Figure 4–19. In connection with the creation of a small boat basin in Santa Cruz, California, a jetty was built in the 1950s to protect the area from strong Pacific swells. Since the building of the jetty, however, longshore sediment transport has been interrupted. Sediments are trapped against the jetty on the upcurrent side, forming a wide beach. Longshore currents now starved of sediments have eroded the coast on the downcurrent side of the jetty, in some cases undermining houses built on the cliffs above.

4-6 Wind and Ice

Water is the dominant shaper of landscapes, except where there isn't any! Thus by default, in deserts, at the poles, or at high altitudes, wind and ice reign as the principal agents of surface erosion and sediment trans-

Figure 4–19

Aerial photograph of Santa Cruz yacht harbor. The jetty interrupts longshore sediment transport so that sediments on the west side of the jetty have collected and formed a wide beach. Longshore currents on the other side of the jetty, now starved of sediments, have eroded from the coast on the east side. To compensate for this erosion, sediments dredged from the harbor entrance are dumped along the beach east of the jetty.

portation. In these places, water is scarce either because rainfall is limited or because it is frozen into ice. Regions where wind and ice assume special importance as surficial phenomena are sparsely populated, however, and so for our purposes do not require extensive discussion. We should mention, though, that many soils today in the northern hemisphere have developed from sediments deposited by glacial ice. Moreover, in some parts of the world, like the Middle East and southwestern United States, attempts are being made to reclaim and settle desert lands.

Deserts and sand

Almost 15 percent of the earth's land surface is *arid*—has an annual rainfall of less than 25 cm—and temperatures are sufficiently high that evaporation is substantial. In such desert regions, the wind's action in eroding and transporting sediments is as significant as the work done by surface water flow in other climes.

The lack of vegetative cover in deserts keeps rock surfaces and their weathering products constantly exposed. Wind blowing across the ground surface picks up the finer-grained surface debris and transports it downwind. As sand grains hit and bounce along the ground, they dislodge other grains, and the impact force moves them a short distance further downwind. Over time, the desert floor is covered with a thin pavement of coarser pebbles left behind by the moving wind, and the finer grained sediment is blown away. Downwind the sand accumulates as sand dunes of various sizes and shapes, depending on wind force (Figure 4–20).

Figure 4-20

Sand dunes in Death Valley, California, that have accumulated by the winnowing action of desert wind moving across unvegetated expanses of loose surface debris.

In some parts of the world, excessive grazing or farming have removed the natural vegetative cover. The light, porous soils thus exposed to the wind are easily eroded. During the 1930s in the central United States, for example, droughts combined with heavy farming and grazing allowed serious wind erosion and brought subsequent failure and abandonment of many farms.

Ice and glaciers

In high latitudes and high altitudes, precipitation falls as snow. If the annual accumulation of snow is greater than the annual melting during warmer parts of the year, the snow accumulates into thick ice layers. Eventually, the ice may be thick enough—several tens of meters—that it creeps slowly outward as a *glacier. Alpine glaciers* form in high altitude mountains and move down their slopes, similar to water courses with their branching tributaries and confined channels. *Continental glaciers,* on the other hand, form continuous ice sheets in high latitudes and cover large parts of continents, flowing radially outward from their centers. They are so thick that they cover mountains as well as valleys. The glaciers that fill some of the valleys and slopes of the Canadian Rockies are examples of alpine glaciers; the Antarctic and Greenland ice caps are examples of continental glaciers.

As just mentioned, ice may pile up to a point where it begins to flow outward very slowly as a viscous liquid. Such behavior is similar to that of rocks within the crust and mantle which flow and exhibit plastic, rather than brittle, behavior. Once again, silly putty makes a good analogy. A pile of silly putty on a table, like a continental glacier, slowly settles into a flat puddle by flowing radially outward. (As they move forward, glaciers gather soil and rock debris at their base and grind this material into finer sediment along their advancing path.)

When glaciers reach lower latitudes or altitudes where temperatures are warmer, the ice melts as fast as the glacier advances, thus bringing on an equilibrium state in which forward movement is halted. Fluctuations in precipitation and temperature cause a glacier to advance or melt backward from its temporary equilibrium position. Melting water at the front of the glacier washes out sediment in the glacier, and this process creates sandy, poorly sorted sedimentary deposits called *glacial outwash*; see Figure 4–21. As the glacier front retreats during a warming trend, the glacial outwash is deposited across the countryside as a thin veneer of sediment. Much of the sand and gravel used for construction comes from these glacially derived sediments. With time, these deposits weather and become material for soils.

Figure 4–21

Bouldery glacially deposited sediments, Connecticut. Notice thin soil that has formed on top.

Sometimes winds blowing across large expanses of glacial sediments pick up the finer silt particles and deposit them downwind as weakly consolidated sediments called *loess*. Loess deposits formed during the last glacial ages are common in central Europe, China, and the Mississippi valley, among other places. Many loess sediments make fertile agricultural lands that are, unfortunately, susceptible to wind erosion during droughts. Large areas of the American Midwest, particularly along the Missouri and Mississippi River valleys, are covered with a blanket of loess a meter to several meters thick.

In the Midwestern United States, successive layers of outwash and soil indicate four major ice advances (outwash) and retreats (soil) during Pleistocene time.

Viewpoint   Luna B. Leopold

Luna B. Leopold holds a joint professorial appointment in the Departments of Geology and Geophysics and Landscape Architecture, University of California, Berkeley. Dr. Leopold was chief hydrologist with the U.S. Geological Survey for a number of years and has made important contributions to the field of urban hydrology. In this Viewpoint Dr. Leopold indicates the ways in which land-use changes affect the hydrology of a local area.

Planning Procedures and Hydrologic Variables

A planning document presented to a community for adoption must always be more suggestive than coercive. This is true not only because the planner is unable to foresee the innumerable complications involved in actual development, but also because many intricate alternatives would accomplish generally similar results. The planner is particularly concerned with both the constraints and the opportunities offered by the principal physiographic characteristics of a particular area, especially the location of hillslopes, soils, and streams. The existing land-use pattern and the accompanying distribution of woods and agriculture may slowly change over a period of years. Roads, villages, industries, and other man-made features are more or less permanent and exert their greatest influence on future development, especially through land values.

    Four interrelated but individual land-use changes can affect the hydrology of an area: changes in peak flow characteristics, changes in total runoff, changes in quality of water, and changes in the hydrologic amenities. The hydrologic amenities are what might be called the appearance or the impression that a river's channel and valleys leaves with the observer. Of all land-use changes affecting an area's hydrology, urbanization is by far the most forceful.

Hydrologic amenity

We can measure runoff by number and by characteristics of rise in streamflow. The many rises in flow and accompanying sediment loads control the stream channel's stability. The two principal factors governing flow regimen are the percentage of area made impervious, that is, covered by concrete or some other impermeable material, and the rate at which water travels across the land to stream channels. The former is governed by the type of land use; the latter is governed by the density, size, and characteristics of tributary channels and thus by the available storm sewerage. Stream channels form in response to the discharge, velocity, and sediment load of the stream. When these elements change through land use or other alterations, the stream channels will adjust to accommodate the flows.

The volume of runoff is governed primarily by the area's infiltration characteristics and is related to land slope and soil type as well as to the type of vegetative cover. It is thus directly affected by the percentage of the area covered by roofs, streets, and other impervious surfaces at times of hydrographic rise during storms. The percentage of an area's impervious surfacing decreases markedly the greater the size of the lot. Felton and Lull (1963) estimate that in the Philadelphia area 32 percent of the surface area is impervious on lots averaging 0.2 acre in size, whereas only 8 percent of the surface area is impervious on lots averaging 1.8 acres.

As the volume of runoff from a storm grows larger, the size of flood peak also becomes greater. Runoff volume also affects later amount of discharge because in any series of storms the larger the percentage of direct runoff, the smaller the amount of water available for soil moisture replenishment and for groundwater storage. When the total runoff from a given series of storms is great because of imperviousness, groundwater recharge decreases so that there is less groundwater contribution during times of low discharge, reducing the discharge still more. Thus, increased imperviousness causes greater flood peaks during storm periods and smaller discharges between storms.

The principal effect of land use on sediment comes from the exposure of the soil to storm runoff. This occurs mainly when bare ground is exposed during construction. It is well known that sediment erosion increases with land slope. Sediment yield in urban areas tends to be larger than in unurbanized areas even if there are only small and widely scattered units of unprotected soil in the urban area. These scattered, bare urban areas do in fact produce considerable sediment.

A major effect of urbanization is the introduction of effluent from sewage disposal plants, and often the introduction of raw sewage, into channels. Raw sewage obviously degrades water quality, but even treated effluent contains dissolved minerals not extracted by sewage treatment. These minerals act as nutrients and promote algae and plankton growth in a stream. This growth in turn alters the balance of organisms in the stream.

Land use in all forms affects water quality. Agricultural use raises the amount of nutrients in stream water; these nutrients come from the excretion products of farm animals and from commercial fertilizers. A change from agricultural use to residential use, as in urbanization, usually reduces these types of nutrients. But this tendency is counteracted by the widely scattered pollutants of the city, such as oil and gasoline products, which are carried through the storm sewers to the streams. The net result is generally an adverse effect on water quality. This effect can be measured by the balance and variety of organic life in the stream, by the quantities of dissolved material, and by the bacterial level. Unfortunately, we have very little data describing and comparing quality factors in streams from urban and unurbanized areas.

Finally, the amenity value of the hydrologic environment is especially affected by three factors. The first factor is the stability of the stream channel itself. A channel, which is gradually enlarged owing to increased floods caused by urbanization, tends to have unstable and unvegetated banks, scoured or muddy channel beds, and unusual debris accumulations. These effects all lower the amenity value of a stream.

The second factor is the accumulation of civilization's artifacts in the channel and on the flood plain: beer cans, oil drums, bits of lumber, concrete, wire—the whole gamut of rubbish from an urban area. Though this debris may not strongly affect the channel's hydrologic function, it deters from what we have been calling the hydrologic amenity.

The third factor affecting the amenity value of the hydrologic environment is the change brought on by the disruption of balance in the stream's wildlife and plants. The addition of nutrients promotes the growth of plankton and algae. A clear stream, then, may change to one in which rocks are covered with slime; in addition, turbidity usually increases, and odors may develop. As a result of greater turbidity and reduced oxygen content, desirable game fish give way to less desirable species. Although lack of quantitative objective data on the balance of stream biota is often a handicap to any meaningful and complete evaluation of the effects of urbanization, qualitative observations tend to confirm these conclusions.

Clearly, planning procedures for land use require knowledge of the hydrologic factors in the local area. The suitability of alternative planning patterns can then be evaluated in hydrologic terms. Planners must therefore consider geologic processes, in this case local hydrology, as well as social, economic, and political questions.

Summary

Solid rocks exposed at the earth's surface are chemically altered and physically disintegrated by weathering processes. The solid and loose rock debris generated by weathering move downhill by mass wasting, pulled by the force of gravity. The downslope processes of creep, solifluction, slumps, slides, mudflows, and rock falls may be fast or slow, wet or dry.

Precipitation also carries loose sediments downslope, when rain and ice flow from higher to lower elevations. Surface water carries these loose sediments in solution, suspension, and by traction down hillsides, emptying them into stream and river channels that flow down valleys and to the sea. Waves and surf attack the land and transport both the sediments formed along the coast by weathering and mass wasting and those carried to the coast by streams and rivers. Streams and rivers, cliffs, beaches, and offshore sand bars all record the work of surface water.

Wind blowing across the arid, treeless deserts sifts sand from the surface layer of rocky rubble and collects it downwind as dunes. Occasional cloudbursts briefly flood the mountain canyons and transport coarse cobbles and pebbles toward the valley below, at whose margin the sediments are deposited as broad, gently dipping alluvial fans.

Glacial ice grinds over the land, scouring the underlying rock and carrying its load of sediment along the glacier's path. Meltwater in the lower reaches of the glacier washes out much of the sediment and deposits it beyond the ice sheet. When the glacier is finally gone, the landscape is strewn with glacial debris collected from kilometers around.

Surficial processes, therefore, work continually and constantly to erode the land and move eroded materials to the sea. With all this wear and tear, it might seem that the continents soon would be leveled—that is, within thousands of years. But the internal processes of orogeny and isostasy, of plate movements and crustal upwarping, constantly rejuvenate the land. These internal processes restore former elevations and thus counteract leveling by surface processes.

Glossary

aggradation Building up of the land surface or sea floor by the accumulation of sediment. Applies especially to stream and river beds that build up to increase their gradient.

aquifer Permeable sediment or rock layer that transports underground water.

capacity The amount of sediment a stream or river can carry past a certain point over an interval of time.

competence Maximum size of sediment particles that a stream or river can carry at a given velocity.

degradation   Lowering of the land surface or sea floor by erosion of sediment or rock. Applies especially to stream and river beds that erode to decrease their gradient.

discharge   Volume of flow of a stream or river past a certain point over an interval of time.

gradient   The slope of a stream or river channel along its length.

hydrologic cycle   Interconnected processes that move water throughout the earth's hydrosphere, including evaporation, precipitation, and surface and underground flow.

load   Amount of sediment that a stream or river carries in solution, suspension, and along its channel by traction.

mass wasting   Downslope movement of surface soil, sediment, and rock by gravity; includes soil creep, solifluction, slides, slumps, rock falls, and mudflows depending on rate of movement and water content of materials.

weathering   Chemical alteration and physical disintegration of earth materials in contact with air, water, and organisms.

Reading Further

Bloom, A. L. 1969. *The Surface of the Earth.* Englewood Cliffs, N.J.: Prentice-Hall. A brief and concise discussion of surficial processes, including rock weathering, mass wasting, streams and rivers, landscape evolution, coastlines, and glaciers.

Garner, H. F. 1974. *The Origin of Landscapes.* New York: Oxford University Press. An advanced text reflecting current theories of how the land is shaped by surface processes.

Gordon, R. B. 1972. *Physics of the Earth.* New York: Holt, Rinehart and Winston.

Leopold, L. B. 1974. *Water, a Primer.* San Francisco: W. H. Freeman. Clear, well-written review of the hydrologic cycle, surface runoff, the work of rivers, and floods.

The Biosphere 5

In exploring the earth's solid and fluid shells we have dwelt on chemical or physical events, usually overlooking living things that compose the biosphere. In a way, life is superfluous to the planet—more an afterthought than a necessity. During the first third of the planet's history, about 1.5 billion years, no life of any kind existed. For almost the next 2.5 billion years, what life there was consisted of primitive plants like algae, fungi, and bacterialike forms. For only slightly more than the last half billion years have higher plants and animals occupied earth, as you can see in Figure 5–1. And the time will come when life will end because the sun will expand and its hot gases will envelope the earth, leaving only the solid shells.

Large-scale geologic processes operate independently of the organisms that have appeared in the last several billion years. The movement of plates, eruption of volcanoes, building of mountains, shaking by earthquakes, and weathering of the land continue according to the laws that govern them, and the presence of living organisms in no way affects their behavior.

But at other scales of time and space, life *has* had an influence on the earth's workings. Microorganisms forming soils, marine shells depositing limestone, plants generating oxygen, and organic matter releasing carbon dioxide are just a few ways that life affects our planet. Humans in particular have their impact: they dump wastes into surface waters; strip natural vegetation; mine and quarry; and scatter radioactive materials. All these activities interfere with natural processes.

No matter how fleeting life in general—and human life especially—may be in the earth's long history, we do see ourselves as custodians of our own lives and other life around us. For this reason, we should understand some of the earth's biological history and processes, so that we may become responsible caretakers for the many and diverse living forms of the biosphere.

Some races wax and others wane, and in a short space the tribes of living things are changed, and like runners hand on the torch of life.

Lucretius, c. 65 B.C.

5-1 Evolution of Life

Not only has the earth changed and evolved through time as a result of chemical and physical processes, but life itself has experienced numerous alterations. From the primitive atmosphere rich in methane, ammonia, carbon dioxide, and water vapor, a variety of complex organic molecules formed in the sea. These molecules eventually led to the earliest life forms about 3 billion years ago; these resembled single-celled algae and bacteria. About this time, too, the crucial biological process of plant photosynthesis evolved. Through ever-increasing photosynthetic activity, and the upper atmosphere's separation of water vapor into hydrogen and oxygen molecules by high energy radiation, the earth's primal atmosphere gradually changed. Eventually, it evolved into mainly nitrogen and oxygen plus small

We discussed photosynthesis and atmospheric evolution back in Sections 1–2 and 1–3.

Figure 5–1

Times of first appearance of major components of the biosphere. Note the relatively late appearance and evolution of animal life during the last one-eighth of earth history.

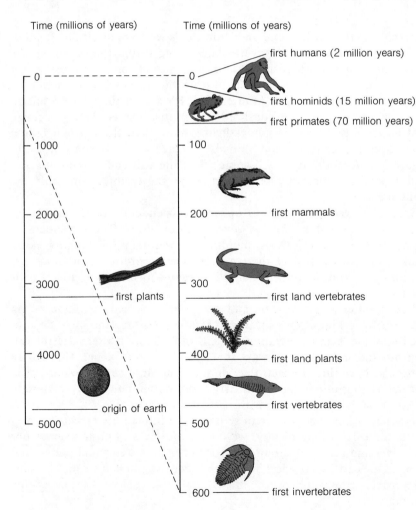

amounts of carbon dioxide and traces of the *noble gases* (gases like helium and argon, which seldom combine with other elements). This atmospheric evolution probably took 2 or 3 billion years.

More than a half billion years ago, higher, or more complex, invertebrates first appeared in the sea. These were followed about 150 million years later by primitive fish. As more oxygen entered the atmosphere, a layer of *ozone* (a gas with three atoms of oxygen bound together) formed in its upper region. This ozone layer blocks out ultraviolet radiation from the sun that is harmful to life. In the sea, such radiation penetrates only a few meters; thus it did not affect life in the early oceans. But until the ozone layer had developed, life could not develop on land and was restricted to water.

About 400 million years ago, the protective ozone layer developed sufficiently for life to invade dry land. The first invaders were plants and insects; they were soon followed by amphibians, and from them arose reptiles. About 200 million years after life developed on land, mammals appeared, gradually developing into such diverse types as shrews, bats, whales, horses, lions, rabbits, and monkeys. About 2 million years ago, or one-twentieth of 1 percent of all earth history, our African ancestors, the proto-human *australopithecines*, appeared.

To visualize how short a time we humans have occupied the earth, convert the evolutionary milestones shown in Figure 5-1 to a 24-hour time scale.

Colonizing new frontiers

Much of the diversity and richness in the unfolding history of life reflects the migration of organisms from marine to terrestrial environments. And within terrestrial environments life spread out: from swampy, coastal lowlands to interior plains and mountains; from forests to deserts; from a ground-dwelling existence to life in trees and air.

Life has thus undergone tremendous expansion by colonizing a myriad of physical environments. And when new environments evolved, even more opportunities for life arose. For example, the evolution of grasses which covered semi-arid plains and plateaus opened a new way of life for grazing, hoofed herbivores such as horses and cattle.

As these two expansions of life took place—into new physical environments and into life forms—a third trend developed. By becoming more specialized within an environment, more species could occupy that environment. Terrestrial mammals proliferated not only by adapting to a wide variety of habitats such as forests, deserts, mountains, tundra, and grasslands, but also by subdividing a habitat's food supply. They achieved the latter by becoming herbivores (cattle), carnivores (cats), scavengers (hyenas), omnivores (primates), grain-eaters (rodents), and insectivores (shrews). Further diversity of species within any one of these feeding types was accomplished by specializing on food of a particular size or variety. For instance, some herbivores graze on grasses while others browse on leafy vegetation.

Enter humans

Our own early ancestors, the australopithecines, appeared roughly 2 million years ago in Africa. They had evolved from a line of ground-dwelling, vegetarian, apelike creatures. The major distinction between early proto-humans and their primate predecessors was the shift to hunting small game in open grasslands. Proto-humans were able to occupy this ecologic niche by modifying their behavior, mostly by using tools for capturing and killing game.

For several hundred thousand years, early humans slowly improved their adaptation as hunters by developing more elaborate cultural behavior; this included socialization, prolonged rearing of offspring, verbal communication, and tool-making technology. Cultural evolution was accompanied by changes in brain size and in the structure of the face and hands. This stage of man-the-hunter and woman-the-gatherer lasted until the end of the last ice age (10,000–12,000 years ago); see Figure 5–2.

During most of this time, humans were well integrated with their environment. Population was small—probably less than a few million persons—and stable. Hunting did not significantly threaten animal populations, although toward the end of this period, when hunting techniques had greatly improved, some human cultures may have wiped out some species of larger mammals. Other cultures of this time hunted solitary game in the forest or caught fish, but this caused little damage to the environment.

About 8000 B.C., after the last ice age, many cultures discovered crop cultivation and animal domestication. These practices changed human history. From them came large and reliable sources of food that permitted rapid population increase. Also, now that there was surplus food on hand, people had time to become full-time artisans, soldiers, scholars, and political and religious leaders. The eventual rise of cities and civilization also depended on food surpluses.

To cultivate plants and raise livestock, humans felled forests to clear the land, dammed streams and rivers for irrigation, and altered the natural environment in many other ways. Humans were now having, for the first time in their history, a significant impact on the earth's environment.

The fossil record

We can interpret the evolution of life forms by digging for traces or actual remains of prehistoric life, called *fossils*. These remains are especially abundant within sedimentary rocks and include tracks, trails, bones, shells, impressions, and wood like those pictured in Figure 5–3. Fossils are not found in igneous or metamorphic rocks because the temperature and pressures at which these rocks crystallize destroy any organism that might be caught within them. Animals and plants with mineralized skeletons (shells, teeth, or bone) or chemically resistant tissue (horn, wood, or

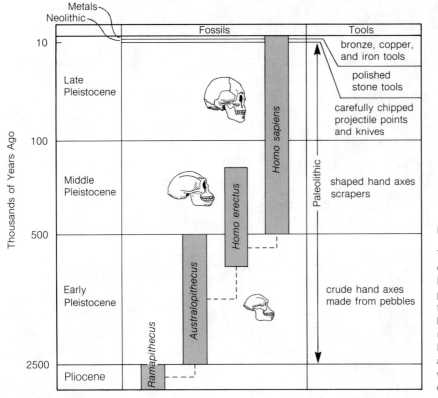

Figure 5–2

The evolutionary stages of humans during the last 2.5 million years. Note the changes in skull shape, particularly the flattening of the face and the increase in relative brain cases. Cultural evolution as recorded by tools is also shown. Paleolithic humans were hunters and gatherers; Neolithic humans were cultivators of crops and domesticators of animals.

cartilage) fossilize more readily than those without such hard parts. Hence, the fossil record of clams or horses is much more complete than that of jellyfish or butterflies. Organisms living in the sea are likely to be fossilized because they live in an environment where burial by sediments occurs continuously. Plants and animals living on dry land are not fossilized as readily, for their remains may be destroyed by predators or erosion before they are buried under accumulating sediments.

Even though fossils enable us to trace the history of life on earth, they are dead and therefore give an extremely limited idea of *how* life evolved. To study the dynamics of evolution, in contrast to its history, we must see how living organisms cope with their environment in surviving, reproducing, and adapting. It is the present-day fauna and flora that give us insights as to how life on earth evolved.

5–2 Evolutionary Processes

Every individual organism, whether a seaweed, fern, worm, beetle, butterfly, starfish, or hippopotamus, has many activities in common with all other

(a)

(b)

(c)

Figure 5–3

Selected fossils. (a) Shelly remains of invertebrates; (b) leaf impression; (c) bones and impression of soft tissues of a marine reptile, called an ichthyosaur.

organisms. Being born, growing and maturing, reproducing, adjusting to the environment, and finally dying are fundamental life processes shared by all animals and plants. The details of individual existence vary tremendously, but these fundamental processes are common to all life. Moreover, their operation over billions of years has filled the earth with a great variety of life. We want now to turn to the evolutionary process and its operation through individual species.

Species: An exclusive group

No organism lives forever. Instead, the thread of life passes from one organism to another either by *asexual reproduction* (in which new individuals bud from a single parent), or by *sexual reproduction* (in which eggs of the female parent are fertilized by sperm or pollen from the male parent). After a period of growth and development, these new organisms become mature and reproduce once again. Eventually, the organism dies by accident or disease, by being preyed upon, or, more rarely, by aging.

Interbreeding among individuals is a crucial aspect of *species:* they are defined as groups of organisms that have the potential to interbreed and produce fertile offspring, and that are unable to do so with other kinds of organisms. This reproductive test measures the similarity of organisms as demonstrated by the genetic compatibility necessary for successful mating. For example, a horse is too genetically different from a dog to produce any offspring. Some organisms may be similar enough genetically that they can mate, but the further requirement of fertile offspring must be met. For example, donkeys and horses can mate and produce a mule, but the mule is sterile. The conclusion is that horses and donkeys are related but separate species.

Members of a particular animal or plant species, being more similar genetically to each other than to members of other species, tend to follow similar ways of life and to fill a distinctive niche in the world's ecology. The concept of the *ecologic niche* has been defined as the way a group of organisms makes its living in nature. Just as there is a biological unity among individuals of the same species in terms of genetic similarity and reproductive compatability, there is also an ecologic unity among these individuals in terms of their environmental needs and adaptations.

Some ecologic niches may be broad, such as that of a bobcat ranging through mountains and valleys to hunt different kinds of prey. Other niches may be more restricted, like that of the Yucca moth, which lives with and feeds on only the Yucca tree. The narrower the niche, the more specialized the way of life of the occupant species. The advantage of specialization is the reduction of competition from organisms in overlapping, adjacent niches; the disadvantage is that even slight environmental change might make the species obsolescent. Some organisms, like the raccoon, fill broad niches that overlap with other species and thereby become competitive. But being less specialized, they are more likely to survive a fairly significant environmental shift.

Survival through adaptation

Every organism inherits from its parents, through the transmission of genes contained in eggs and sperm, a program of instructions enabling the organism to grow, feed, avoid enemies, nest, and mate. These genetic instructions may be precise and detailed, in which case the organism's behavior is largely predetermined. The term "instinct" loosely describes this rather rigid behavior inherited from parents. Other organisms inherit more generalized genetic instructions; their behavior is more flexible and can be modified to particular environmental conditions or situations. Humans are perhaps an extreme example of this latter sort of genetic programming.

Even though we inherit many physical, mental, psychological, and behavioral attributes from our parents, we still have considerable margin for modifying the degree to which our attributes are exercised.

The ability to adapt to a variety of environmental situations is, of course, an adaptation in itself.

An organism's ability to cope with existence in an environment measures the organism's adaptation to that environment. Every sexually reproduced organism is essentially unique in its inherited genetic makeup. Even offspring of the same parents do not have exactly the same genetic constitution. The reason for this diversity is that genes are shuffled during the production of eggs and sperm and during their union, producing new organisms with unique gene blends. Individual differences in experience and learned behavior, along with variations in early growth and development, cinch this uniqueness. The result is that every individual, no matter how apparently similar, differs from every other and therefore conducts its life differently from others. Differences among organisms may be expressed in obtaining food more efficiently, detecting a predator sooner, resisting disease, or making a shelter that is safer, warmer, or drier—in other words, in adapting better to the environment. If the ability to adapt better is genetically determined, and if an organism reproduces more offspring than other individuals in the same niche—because of its more favorable adaptations—then the genes controlling or influencing adaptation in that niche will proportionally increase in the next generation.

The principle of *natural selection*, which we have just described, is the basis for the theory of organic evolution developed by the great, nineteenth-century British naturalist, Charles Darwin. Darwin's theory consists of four simple, basic ideas:

- There is natural variation within a species.
- A species produces more offspring than will survive to reproduce.
- Some offspring are better able to adapt to their environment than others.
- The organisms that adapt better reproduce more and pass on favorable adaptations to their offspring.

These evolutionary processes explain why there is more than just one kind of animal or plant.

The species thus evolves over time by the cumulative effect of many small changes; for a specific example, see Figure 5–4.

5-3 The Environment

The life and death of an organism and its adaptation and selection take place in the surrounding environment. In simplifying our discussion in the preceding section, we mentioned adaptations only in terms of feeding, avoiding predators, seeking shelter, and mating. But organisms adapt to their environments in many other ways. A host of physical, chemical, and biological factors confront plants and animals, providing challenges that determine the organisms' distribution and abundance in nature. This is the subject of *ecology*, namely, the interaction of animals and plants

Figure 5–4

Evolution of the long-necked animals we call giraffes. Presumably, there was variation in neck length in early giraffe populations. Those animals that could reach higher into trees and exploit an abundant food resource were more successful in rearing offspring than those that were shorter and forced to compete with other smaller plant-eaters. The genetic information resulting in taller animals was passed on to these more numerous offspring who, in turn, were favored by being taller, and they produced more offspring.

with their physical, chemical, and biological environment. The two fundamental ecological questions are: what organisms occur where? and how come?

Physical limits to life

Certain physical conditions limit the abundance and distribution of particular kinds of organisms. Factors like sunlight, temperature, moisture, and soil are essential for thriving plants. For terrestrial animals, temperature, humidity, and shelter are critical. And for freshwater or marine animals, existence depends on temperature, depth, water turbulence and turbidity, and bottom sediments.

Plants need solar radiation and moisture to photosynthetically convert carbon dioxide and water into food. Marine plants—almost all of which are the primitive algae—flourish in the upper 200 meters of the sea because sunlight does not penetrate below this depth. Most marine photosynthesis is carried out at depths much shallower than 200 meters by many minute, floating plants called *phytoplankton*. These tiny plants provide the first crucial step in marine food production.

Terrestrial plants, including mosses, ferns, pines, grasses, and hardwood trees, make special demands on the soil in which they are rooted. They require water; nutrient elements like nitrogen, phosphorus, and potassium; and the proper soil grain size and texture. For example, cacti prefer dry, sandy, shallow soils, whereas redwoods like moist, deep, loamy soils. Terrestrial plants are the first step in food production on land.

Plants are sensitive also to temperature; both marine and land plants vary systematically with geographic latitude and topographic altitude. The earth displays zones of plant types from colder, high latitudes and altitudes

Do animals show a similar latitudinal and altitudinal zonation?

to warmer, low latitudes and altitudes, as indicated in Figure 5–5. Terrestrial animals, too, are influenced by surrounding temperature; most animals' body temperatures fluctuate with the external temperature. (Mammals and birds are exceptions. They maintain a constant internal temperature despite the external temperature—within limits, of course.) As external temperature rises or falls, so do an animal's rates of metabolism, development, growth, and reproduction. These biological processes are driven by chemical reactions whose rates depend on temperature.

Temperature has other, less direct influences on organisms. The solubility of solids and gases in water—whether seawater, lakes, streams, or soil moisture—is temperature-dependent, and thus their concentration varies with temperature. As an example, gaseous oxygen in water solution escapes to the atmosphere with an increase in temperature. A river receiving hot water from a power plant or factory may have much of its normal oxygen reduced by the rise in water temperature. Consequently, the river's organisms, for instance fish and invertebrates, may be asphyxiated.

Temperature also controls the density of air and water masses. Hotter or colder temperatures lower or raise their density and thereby force the

(a)

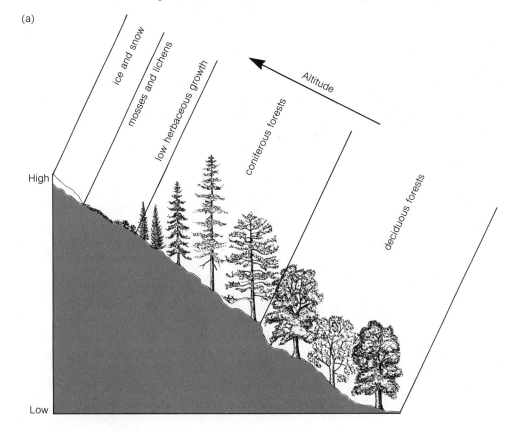

water or air mass to rise or sink. These density changes create convection currents in the atmosphere and hydrosphere, and help keep biologically vital materials, like oxygen and nutrients, in circulation.

Turbulence and turbidity of their water habitat has a definite impact on aquatic organisms. Water that is too turbulent can dislodge animals and plants growing on the bottom, bury them with sediment, or even injure them by its force. Water that is too quiet may not circulate enough to allow adequate distribution of oxygen or essential nutrients. Intermediate levels of turbulence are desirable because they keep food and oxygen moving around the local environment. Overly turbid water has so much sediment or organic matter in suspension that it clogs the feeding apparatus of animals, buries them when it settles, or shuts out light and prevents photosynthesis.

Many aquatic organisms live on river and lake bottoms or on the sea floor, and these organisms prefer various kinds of sediment substrates. Organisms actually attached to the bottom—like seaweeds, sponges, corals, and barnacles—prefer hard rock or firm, sandy substrates. Unattached animals that burrow through the substrate for food or shelter—like marine worms, some snails, and many clams—prefer soft, muddy sediments.

Figure 5-5

Latitudinal variations in plant types are paralleled by similar variations in altitude. In both cases, the chief environmental factor controlling the distribution and abundance of the plants is temperature.

(b)

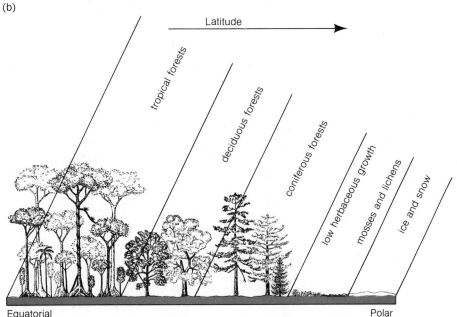

Living through chemistry

Just as living organisms require certain physical conditions, so do they also have basic chemical requirements. They need nutrient elements to consume, carbon dioxide to photosynthesize or oxygen to breathe, and, for aquatic organisms especially, proper amounts of dissolved substances in the surrounding water. In addition to water and carbon dioxide for photosynthesis, plants must have small amounts of essential elements like nitrogen, phosphorous, and potassium. Animals, too, depend on these and other elements such as iron, calcium, iodine, and magnesium to support life processes. When animals and plants *metabolize* their food (that is, break it down chemically to release its stored energy and nutrients), they use oxygen to accomplish this process. Even aquatic animals depend on oxygen dissolved in local water, as do soil organisms like earthworms and insect larvae, which take gaseous oxygen from the interstices of soil grains.

Water continually comes into contact with fluid-filled organic tissues: water is drunk by terrestrial animals; dew and soil moisture sustain plants, and fresh or marine water bathes aquatic organisms. Since water is so essential to life, the amount of dissolved substances contained in it is critical for practically all organisms. If too many or too few dissolved salts are in the external water as compared to the cell fluids, its salinity will influence how much water organisms lose from, or add to, their tissues by *osmosis*. That is, water will flow through cell walls from a less salty fluid to fluids with more salt until the salt concentration in both fluids is about equal. If too much water moves into cells, they become flooded and eventually burst; if too much water moves out of cells, they lose too much water and collapse (Figure 5–6).

The crinkling of your fingers when you've had your hands in water for a long time is an osmotic phenomenon . . . so is the wilting of salad lettuce when not served quickly enough. Why?

Thus most terrestrial animals cannot drink saline water because its high salt content dangerously dehydrates body tissues. Nor can most freshwater fish and invertebrates survive long in marine waters, where the salinity is almost 100 times greater than fresh water. Conversely, most marine organisms, adapted to higher salinities of the seas, cannot live in fresh water. The reason is that the salt concentration in their body cells is so much higher than that of the fresh water that the latter floods their body cells and ruptures them. Some marine organisms, like sharks, salmon, eels, and clams, can tolerate somewhat lower than normal marine salinity by regulating the flow of water that enters their tissues. Although these organisms usually live in the ocean, they can penetrate estuaries or even freshwater environments.

Biological boundaries

Organisms limit each other in their distribution and abundance. The principal source of interaction among living things is food-gathering in which one organism feeds on another. The abundance of a particular

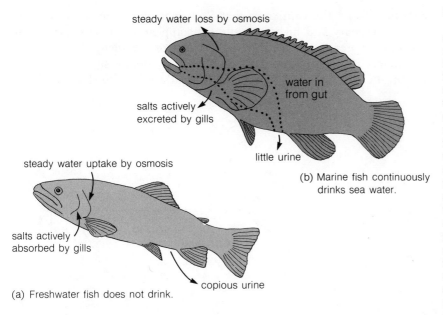

steady water loss by osmosis

water in from gut

salts actively excreted by gills

little urine

(b) Marine fish continuously drinks sea water.

steady water uptake by osmosis

salts actively absorbed by gills

copious urine

(a) Freshwater fish does not drink.

Figure 5–6

Adaptations in freshwater and marine fish for maintaining the balance of water and salts in body tissues. (a) In freshwater fish the salt concentration of the body fluids is greater than in the surrounding water. To prevent too much dilution of the body fluids these fish remove the excess water in their urine; some body salts are also lost in the urine, so the fish replace the salts by absorbing them from the external water. (b) Marine fish have the opposite problem in that they lose body water to the saltier surrounding water. To maintain their water-salt balance these fish excrete little urine, absorb seawater through the gut, and remove excess salts in the seawater by the gills.

animal is controlled by the availability of its food supply. Plants, which make their own food, are fed upon by the plant-eating herbivores that are, in turn, fed upon by the meat-eating carnivores. Thus, arctic lichens feed the moose that feed the wolves. Phytoplankton in the sea feed small, floating animals called *zooplankton* that feed fish, like herring, that feed larger fish, like sharks. Seed- and fruit-eating finches and sparrows are eaten by hawks, owls, and maybe your cat. In short, the distribution and abundance of predators are limited by the distribution and numbers of their prey, who are in turn controlled by the presence and abundance of their prey (if carnivores) or plants (if herbivores). Other animals, like some worms, crabs, snails, and fish in the sea, and many terrestrial insect larvae, are *scavengers*—they feed on dead animal and plant matter.

In aquatic environments, many invertebrates feed by filtering suspended organic matter and plankton from the surrounding water. Sponges, corals, oysters and other clams, barnacles, and the baleen whale are examples of marine *filter feeders*. Other aquatic organisms, like sea urchins, worms, some clams and snails, feed on the organic matter in bottom sediments, and are known as *deposit feeders*.

Great variety of feeding types among animals reflects the effective and efficient utilization of an environment's food resources. Further subdivisions within a feeding type represent specializations for one particular food source. For example, predators like bobcats, owls, and snakes feed on different kinds of herbivores.

Plants, remember, are the first step in food production and therefore are the primary producers in the environment. Herbivores are primary consumers, and the carnivores that feed upon them are secondary consumers. Tertiary and higher consumers complete the *food chain:* plants ⟶ herbivores ⟶ carnivores ⟶ carnivores. Usually the diet of the consumers is sufficiently varied that there are many such food chains within the environment; together they form a *food web* (Figure 5–7).

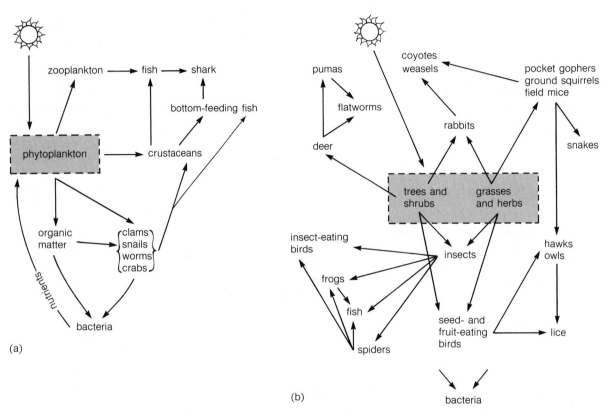

Figure 5–7

Multiple food chains forming food webs in a marine (a) and terrestrial woodland (b) environment. In the marine environment, the small floating plants (phytoplankton) feed the small floating and swimming animals (zooplankton) who, in turn, feed crustaceans, sharks, and bottom-feeding fish, like the cod. Living and dead organic matter in the water and on the sea floor feeds clams, snails, worms, and crabs. Bacterial decay of the unconsumed organic matter releases nutrients to the surrounding water. Similar feeding relationships exist in the terrestrial woodland environment. Plants feed a variety of herbivores, like deer, rabbits, rodents, insects, birds, frogs, and fish. Predators, such as pumas, weasels, coyotes, snakes, hawks, owls, and fish, feed on these herbivores. Scavengers and bacteria feed on the accumulated organic from all these organisms.

128

Relations among environmental controls

Not all physical, chemical, and biological conditions in an environment are equally significant in determining the success of the living inhabitants. Usually one factor, the so-called *limiting factor*, exerts the greatest influence in a given environment. For example, the availability of nitrogen in soils is a common limiting factor in plant growth; that is, some plants are most dependent on this one environmental variable. In other environments water may be the single most important factor in the growth of plants and colonization of animals.

What is the limiting factor for humans?

We should also note that the physical, chemical, and biological conditions of an environment interact with each other. Few, if any, operate independently of the other. Consider again streams or lakes heated by water from a power plant or factory. The higher temperature raises respiration rates, so organisms need more oxygen. Unfortunately, increased temperature also releases gaseous oxygen from water. Thus, animals residing in rivers heated by industrial effluents need more oxygen, but they cannot get very much because heating releases some of it. If the effluent also contains organic matter, some of the available oxygen will oxidize it, which makes still less oxygen available to the organisms.

Choosing a place to live

Anything more than a casual look at the natural world reveals that its physical, chemical, and biological attributes vary greatly. Not only are there obvious differences among forests, deserts, grasslands, seashores, coral reefs, and arctic ice, but variations exist within a homogeneous environment. For example, grasslands are not a continuous carpet of wild oats. A grassland environment like that in Figure 5–8 may accommodate other grasses, wild flowers, scattered clusters of trees, hills, ravines, rock out-croppings, and an occasional spring or stream whose banks are lined with cottonwoods, alders, and willows. Variations in shade and exposure to sunlight, soil moisture, vegetative cover, and so forth result in many microenvironments within the grassland that support an equal variety of animal and plant life. Geographic differences also enter the picture. Grasslands of southern California are, among other things, drier and hotter than those in northern California, some 1000 kilometers away. Or consider the New England coastline: salt marshes, bays, estuaries, sandy beaches, and rocky promontories. Such geographic features, plus differences in water salinity, turbulence and turbidity and sediment substrates, create a kaleidoscope of environments, as you can see in Figure 5–9.

Variations in the environment influence the dispersal, distribution, and density of species. Within the overall geographic range of a species—for example, the California grasslands stretching from the Oregon border to Mexico or the New England coast from Maine that faces the Atlantic to Connecticut that fronts along Long Island Sound—some areas are more favorable to a given species than others. A species distributes itself accordingly throughout its range. Depending on an environment's heterogeneity, organisms spread themselves continuously throughout the range or discontinuously group together in small clusters in areas where conditions are more favorable. Deer live mainly on the wooded stream banks and ravines of California's grasslands; mussels cluster on intertidal rocks along New England's coast. Other areas within California's grasslands or New England's coast have few, if any, deer or mussels.

Seasonal or longer-term environmental changes may force a species to contract or expand its range, and consequently increase or decrease its density within the range. The terrestrial, hoofed herbivores of the East African Serengeti Plain, like zebra, wildebeest, gazelle, gnu, topi, and impala, migrate more than 1600 km each year, following the rainfall pattern and seasonally available vegetation.

Figure 5–8

Environmental variation within a California grassland. In addition to open, grassy areas, there are scattered groves of oaks, streams running through ravines lined by cottonwoods, alders, and willows. Differences in exposure to sunlight, shade, and availability of water cause differences in vegetation and animals.

5–4   Atmosphere and Climate

What we call *climate* encompasses environmental conditions like temperatures, humidity, wind, sunshine, cloudiness, and rainfall, averaged over one or several years. *Weather* refers to the varied state of these conditions at any given moment. The environmental conditions comprising weather and climate are determined by the heating of the atmosphere by solar radiation.

Solar heating

Heat energy emitted from the sun has been relatively constant over the eons since the earth was formed, but the amount received varies over the globe. Beyond differences from day to night, incoming heat fluctuates from the equator to the poles. While the earth consistently rotates on

Figure 5–9

Environmental variation within the New England coastline. Rocky headlands, protected sandy beaches, and salt marshes in this region vary in bottom sediments, water turbulence and turbidity, and salinity. This causes local differences in kinds and abundance of marine organisms living along the coast.

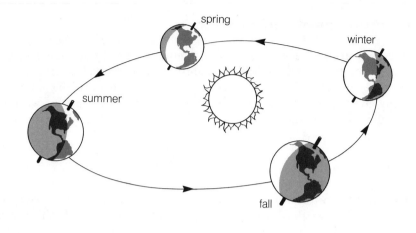

Figure 5–10

Differences in solar radiation during the year. The earth's pole of daily rotation is inclined as the earth revolves about the sun during the year. Consequently, the northern hemisphere receives more radiation in the summer than the winter. Not only are the days longer but the sun's rays are more direct, and hence the amount of radiation received per unit area is greater. During the northern hemisphere summer, of course, the southern hemisphere is experiencing winter.

its tilted axis each day and revolves around the sun each year, polar regions get far less solar radiation than equatorial regions. We can give essentially two reasons for this difference. First, because the earth tilts, the sun's rays fall at a more oblique angle on the poles than the equator, so energy striking the poles spreads over a larger area. Second, these rays travel a longer path through the atmosphere to reach high latitudes than low latitudes. Thus, more energy reaches the lower atmosphere and ground surface near the equator than near the poles; see Figure 5–10. During much of the year, the polar regions do not receive any direct solar energy at all. The middle latitudes also receive less energy at this time, and this is what we call "winter."

The net accumulation of heat makes the atmosphere in equatorial regions warm and less dense. Equatorial air thus rises and drifts toward the poles where it cools and sinks. Cooler, denser polar air flows back to the equator below the warmer equatorial air headed for the poles, as illustrated in Figure 5–11. The rotation of the solid earth beneath its fluid atmospheric shell complicates this simple circulation. As the earth rotates, air shifts eastward on its way to the poles and westward as it returns to the equator (the patterns are reversed in the southern hemisphere); atmospheric circulation is generated in this way.

Air masses

Atmospheric air masses circulating over the earth's surface are characterized by temperature and humidity, or water vapor content. Temperature differences between individual air masses are mostly due to variations in solar heating. As we have just seen, cold air masses typically originate in high latitudes, while warm ones come from equatorial regions. Gaseous water vapor evaporated from land and sea is more readily held by warm air than cold air. When warm air cools—either by mixing or "colliding" with cold air, by moving poleward, or by rising vertically into higher and colder parts of the atmosphere—the water vapor condenses as liquid droplets that form fog or clouds. If enough condensation occurs, the water falls or *precipitates* as hail, snow, or rain.

What we have so far described accounts for a large part of our global climate. Ground elevation, differences in heat absorption and radiation of land and sea, and local wind patterns also determine the temperature and humidity of air masses and their movement. In turn, the character of individual air masses determines environmental conditions like soil moisture, relative rates of evaporation and precipitation, days of sunshine and cloudiness, as well as local air temperature and humidity. And these environmental conditions exert control over the kinds of plants that grow in an environment and the kinds of animals that feed on them.

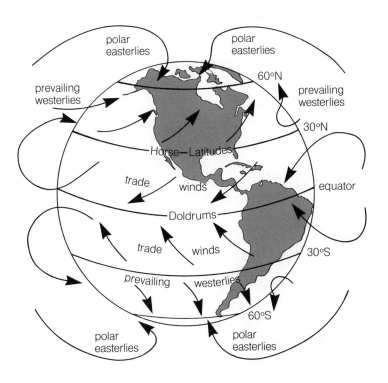

Figure 5–11

Major atmospheric circulation patterns. Air warmed at the equator rises and moves poleward where it cools, sinks, and returns to the equator. This convection pattern is altered by the earth's rotation. As the equatorial air masses move north they are deflected to the right, forming the high-altitude westerly winds in the middle latitudes. The air masses returning from the pole also turn to the right, creating the low-latitude easterly winds near the surface. In the southern hemisphere air masses are deflected to the left, rather than the right.

Climatic conditions vary over the earth's surface and over time and thus influence the spatial and temporal distribution of plants and animals. Moreover, during earth history as the distribution of lands and seas changed, as mountains were uplifted by orogenies or leveled by erosion, and as continents wandered, and as sea level fluctuated, the earth's climates varied considerably. No wonder, then, that the biosphere is so rich in complexity, not only today, but throughout all earth history, from place to place and from time to time.

5-5 Ecosystems

Commoner's Viewpoint at the conclusion of this chapter focuses on the interactions of diverse parts of the ecosystem.

An *ecosystem* is the sum of all interactions of animals and plants with each other and with their physical and chemical environment (Figure 5–12). Solar energy powers the ecosystem and essential nutrients circulate through it. Primary producers, primary, secondary, and tertiary consumers, and decay organisms each play an important role in the functioning of an ecosystem. Simply by knowing whether a particular ecosystem is a lake, mountain stream, salt marsh, or coral reef, we can identify individual species that fill the ecological roles there. The sea's phytoplankton and algae perform the same primary food production role as do grasses and trees in a terrestrial woodland. Sharks are surely not the same kind of

Figure 5–12

Diagram showing the integration of physical, chemical, and biological components of terrestrial and marine ecosystems. Solar energy, trapped by plants as food, drives the system. Nutrients are recycled by decomposer microorganisms that release nitrogen, phosphorus, and other critical substances initially bound up in plant and animal matter. The ecologic roles of producer, consumer, and decomposers are filled by different kinds of animals and plants in different ecosystems.

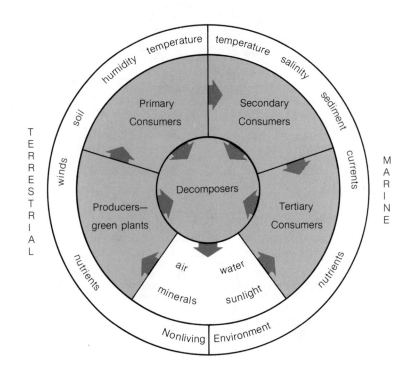

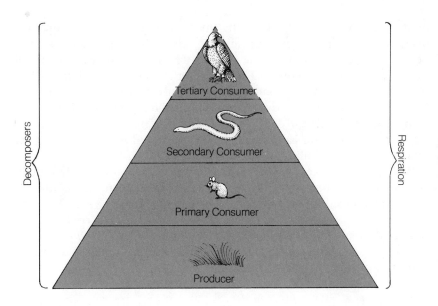

Figure 5–13

Pyramid of energy in an ecosystem. Area of each food level indicates relative amount of energy at each level. Only a part of the energy at any level is passed to the next level because some energy is lost within it by respiration or decomposition of dead organisms.

animal as bobcats, but both are secondary consumers in their own ecosystem.

Energy and nutrients circulate through the ecosystem, so that food produced by plants passes to higher and higher consumer levels. At each level in the food chain only about 10 to 20 percent of the available energy is passed on to the next level as food. Organisms within the same level use the remaining energy for growth, reproduction, and activity. Some energy is also lost to the decay part of the cycle or is radiated as heat to the external environment (Figure 5–13).

Diversity and stability

Ecosystems with many different kinds of organisms at the producer and consumer levels appear to be more stable than those with just a few kinds. The effects of fluctuation in physical or chemical conditions are minimized within a diverse ecosystem having many food webs, as opposed to one with a single food chain. A dry spell, cold wave, epidemic, or other environmental change in a complex ecosystem affects only some component species; other species survive and temporarily grow in abundance. In a simple ecosystem, such environmental changes may kill the one or two species occupying a level in the food chain, thereby toppling the entire system.

On land, tropical ecosystems with mild temperatures and abundant water tend to be complex and stable. As temperatures become colder and water scarcer, ecosystems lose diversity and stability. For example, wild swings in prey population of arctic hare are paralleled by swings in the population of its predator, the arctic lynx. In this simple arctic food chain, hare overpopulation causes epidemic disease, plant shortages, and a scarcity of nesting sites. A sharp decline in hares brings on an equally sharp fall in the population of lynxes. The resulting shortage of predators allows the hare population to boom once again, only to dwindle several years later. The drawbacks of ecosystems with limited food chains were tragically demonstrated in the early 1970s when droughts in central Africa diminished grazing areas for cattle that supplied meat and milk. The result was disastrous famines killing thousands of people.

When humans intervene in natural environments, they usually simplify ecosystems and threaten their stability. Clearing land for agriculture removes the protective cover for natural predators of insect pests. After crops begin to grow on the cleared land, occasionally such pests destroy them. In a similar way, killing local predators to protect domestic fowl and cattle can result in population booms of resident deer, followed by starvation and mass mortalities among the deer.

Viewpoint   Barry Commoner

Barry Commoner is Professor of Botany, Washington University, St. Louis. For many years Dr. Commoner has given public lectures and written articles and books to alert citizens to their impact on the natural environment. In this Viewpoint Dr. Commoner succinctly outlines the four laws of ecology that we must keep in mind if we are to live in balance with animate nature.

The Four Laws of Ecology

The First Law of Ecology:
Everything Is Connected to Everything Else

The single fact that an ecosystem consists of multiple interconnected parts, which act on one another, has some surprising consequences. Consider, for example, the freshwater ecological cycle: fish—organic waste—bacteria of decay—inorganic products—algae—fish. Suppose that unusually warm summer weather speeds up the production of algae. The rapidly growing algae deplete the supply of inorganic nutrients so that two sectors of the cycle, algae and nutrients, are out of balance in opposite directions. The operation of the ecological cycle soon brings the situation back into balance. For the excess in algae makes it easier for fish to feed on these plants; this reduces the algae population, increases fish waste production, and eventually leads to a greater level of nutrients when the waste decays.

Thus, the levels of algae and nutrients tend to return to their original balanced position.

If the entire cyclical system is to remain in balance, the overall turnover rate must be governed by the slowest step—in this case, the growth and metabolism of the fish. Any external effect that forces part of the cycle to operate faster than the overall rate leads to trouble. So, for example, the rate of waste production by fish determines the rate of bacterial decay and the rate of oxygen consumption resulting from that decay. In a balanced situation, the algae gives off enough oxygen into the air to support the decay bacteria. Suppose that the rate at which organic waste enters the cycle is increased artificially, for example, by dumping sewage into the water. Now the decay bacteria are supplied with organic waste at a much higher level than usual; because of their rapid metabolism they are able to act quickly on the increased organic load. As a result, the rate of oxygen consumption by the decay bacteria can easily exceed the rate of oxygen production by the algae (and the oxygen's rate of entry from the air), so that the oxygen level drops to zero and the system collapses. Thus, the rates of the separate processes in the cycle are in a natural state of balance that is maintained only so long as nothing intrudes on the system from the outside. When such an effect originates beyond the cycle, it is not controlled by the self-governing cyclical relations and it therefore threatens the system's stability as a whole.

All this results from a simple fact about ecosystems—everything is connected to everything else. The system is stabilized by its dynamic self-compensating properties; these same properties, if overstressed, can lead to a dramatic collapse; the complexity of the ecological network and its intrinsic turnover rate determine how much interference the network can endure and for how long, without collapsing; the ecological network is an amplifier, so that a small disturbance in one place may have large, distant, long-delayed effects.

The First Law of Ecology

The Second Law of Ecology:
Everything Must Go Somewhere

This is, of course, simply a somewhat informal restatement of a basic law of physics—that matter is indestructible. Applied to ecology, the law emphasizes that in nature there is no such thing as "waste."

A persistent effort to answer the question "Where does it go?" can yield a surprising amount of valuable information about an ecosystem. Consider, for example, the fate of a household item that contains mercury—a substance with serious environmental effects that have just recently surfaced. A dry-cell battery containing mercury is purchased, used to the point of exhaustion, and then "thrown out." But where does it really go? It is placed in a container of rubbish; the rubbish is collected and taken to an incinerator. Here the mercury is heated; this produces mercury vapor, which the incinerator stack emits into the air. Mercury *vapor* is toxic. The wind carries mercury vapor, and eventually rain or snow brings the vapor to the earth. Entering a mountain lake, let us say, the mercury condenses and sinks to the bottom. Here bacteria convert it to methyl mercury. Methyl mercury is soluble and taken up by fish; instead of being metabolized, the mercury accumulates in the organs and flesh of the fish. Someone catches and eats the fish, and the mercury becomes deposited in that person's organs, where it might be harmful. And so on.

This is an effective way to trace out an ecological path. It is also an excellent way to counteract the prevalent notion that something regarded as useless simply "goes away" when it is discarded. Nothing "goes away"; it is simply transferred from place to place, converted from one molecular form to another, acting on the life processes of any organism in which it temporarily lodges.

The Third Law of Ecology:
Nature Knows Best

One of the most pervasive features of modern technology is the notion that technology is intended to "improve on nature"—to provide food, clothing, shelter, and means of communication and expression superior to those available to man in nature. Stated baldly, the third law of ecology holds that any major man-made change in a natural system is likely to be *detrimental* to that system. This is a rather extreme claim; nevertheless I believe it has a good deal of merit if understood in a properly defined context.

One of the striking facts about the chemistry of living systems is that for every organic substance produced by a living organism, there exists, somewhere in nature, an enzyme capable of breaking that substance down. In effect, no organic substance is synthesized unless there is provision for its degradation; recycling is thus enforced. Thus, when a new man-made organic substance is synthesized with a molecular structure that departs significantly from the types that occur in nature, it is probable that no enzyme capable of breaking down the substance exists. The material then tends to accumulate.

Given these considerations, it would be prudent, I believe, to regard every man-made organic chemical *not* found in nature and having a strong action on any one organism as potentially dangerous to other forms of life. This view means that we should treat all man-made organic compounds that are at all active biologically as we do drugs, or rather as we *should* treat drugs—prudently, cautiously.

The Fourth Law of Ecology:
There Is No Such Thing As a Free Lunch

In my experience, this idea has proved so illuminating for environmental problems that I have borrowed it from its original source, economics. The "law" derives from a story that economists like to tell about an oil-rich potentate who decided that his new wealth needed the guidance of economic science. Accordingly he ordered his advisers, on pain of death, to produce a set of volumes containing all the wisdom of economics. When the tomes arrived, the potentate was impatient and again issued an order—to reduce all the knowledge of economics to a single volume. The story goes on in this vein, as such stories will, until the advisers are required, if they are to survive, to reduce the totality of economic science to a single sentence. This is the origin of the "free lunch" law.

In ecology, as in economics, the law is intended to warn that every gain is won at some cost. In a way, this ecological law embodies the previous three laws. Because the global ecosystem is a connected whole not subject to overall improvement and in which nothing can be gained or lost, anything extracted from the system by human effort must be replaced. We cannot avoid paying this price; we can only delay. The present environmental crisis is a warning that we have delayed nearly too long.

Summary

The biosphere forms a thin and tenuous shell around the earth. Energy from the sun and nutrients from the land and sea are circulated and recycled through the many different ecosystems composing the biosphere. Individual animals and plants aggregate into interbreeding species, each of which occupies an ecologic niche and performs an important role in the functioning of these ecosystems. The adaptation of each species to the physical, chemical, and biological attributes of their environment are genetically controlled and passed on to their progeny. Over the generations, natural selection has resulted in the precise adjustment of each species to its environment. The wide variety of animals and plants across the face of the earth reflects the many different ways of life possible within the biosphere.

Organisms which have adapted to a certain environment either expand, diminish, or die out when the environment changes. Such changes have greater impact on simple ecosystems than on complex ones; in the latter the diversity of organisms and their interactions tends to minimize the effects of environmental fluctuations. Human management of or accidental interference with ecosystems usually upsets the system's balance by varying environmental conditions or by reducing the system's complexity and hence stability. This intervention can do irreparable damage to the system.

Life appeared several billion years ago on this planet and has flourished since. Humans are late-comers to the earth, and if they are to survive, they must learn to manage carefully its biological resources; humans also operate within an ecosystem and, like other organisms, are dependent on its well-being.

Glossary

adaptation Adjustment of plants and animals that permits survival and successful rearing of offspring.

ecologic niche Place within the economy of nature that a particular species of plant or animal occupies. Often described as the way an organism makes its living.

ecology Study of the physical, chemical, and biological factors that control the distribution and abundance of organisms.

ecosystem The totality of interacting phenomena—animal, vegetable, and mineral—within a life habitat that determines the numbers and kinds of organisms present. The ecologic system is usually considered closed in that the necessary energy and nutrients are continually recycled within it.

food chain Transfer of energy, in the form of food, from the sun through primary producers (plants) to primary and secondary consumers (animals). *Food webs* are overlapping food chains.

fossil Traces or actual remains of a prehistoric plant or animal found in sediments or sedimentary rocks.

limiting factor That physical, chemical, or biological requirement that is in shortest supply within an ecosystem.

natural selection All the mechanisms within nature by which some adaptations are more successful than others, leading to differential survival and reproduction within a species.

species   Group of organisms whose members are similar to each other and that can interbreed to produce fertile offspring.

Reading Further

Clarke, G. 1965. *Elements of Ecology*, 2nd ed. New York: John Wiley. A primer on the physical, chemical, and biological factors controlling the abundance and distribution of organisms and their integration into ecosystems.

Dasmann, R. 1968. *Environmental Conservation*, 2nd ed. New York: John Wiley. The major ecosystems of the world and how humans manage and mismanage them.

McAlester, A. L. 1968. *The History of Life*. Englewood Cliffs, N.J.: Prentice-Hall. Narrative history of the evolution of plant and animal life during the last half billion years, leading up to our own species.

The Biosphere. 1970. San Francisco: W. H. Freeman.

Severely eroded landscape in Bolivia.

Time and Change   6

The earth is a dynamic planet; that has been our ongoing refrain in the past five chapters. Constant change has marked the earth's 4.5-billion-year history. The early earth's spheres differentiated themselves into core, mantle, and crust; the atmosphere and hydrosphere developed from gases vented by the mantle and crust. Mineral assemblages and rock types are constantly being renewed by igneous, sedimentary, and metamorphic processes. Shifting lithospheric plates continually shape mountains, create volcanoes, generate earthquakes, and move continents. The land is relentlessly weathered and eroded. Animals and plants adapt to and become integral parts of cycling ecosystems. All these processes and events bear witness to the earth's changing faces through time.

In this chapter we try to synthesize our view of basic earth processes in the context of time and change. We remind ourselves that this planet's present state is a product of past processes, that the earth's condition was different in the past, and that it will undoubtedly be different in the future. We also examine the rates at which changes occur. Some, like plate tectonics, happen so slowly that we can assume no significant change in the near future. Others, like coastal erosion, happen fast enough that we should consider them in making plans for environmental design and use. Still other geologic events, like massive eruptions of long dormant volcanoes, are rare, generally unpredictable, and of sufficient magnitude that, while we can't reasonably shape our lives around them, at least we ought to be able to identify them.

The world's a scene of changes, and to be Constant, in Nature were inconstancy.

Abraham Cowley, 1647

6-1 Establishing Geologic Time

Throughout human history people have created legends and myths, expressed in their folklore or religions, to explain the earth's origin. However conflicting these explanations were, they all had one common assumption: once the earth was created, that was that; little or no further change occurred. In the Judeo-Christian tradition, the earth is believed to have been created by God in much the same state as it appears today. Indeed, in the middle of the seventeenth century, an Irish prelate named Archbishop James Ussher worked out the presumed time of origin by using Old Testament chronologies and dated the Creation at 4004 B.C. For many years thereafter, Western theologians accepted not only an essentially instantaneous creation, but one that occurred in the relatively recent past.

Today, ideas of a brief creative event occurring just a few millenia ago seem naive and quaint. But it was only the careful study of earth history, as revealed by the record of rocks, that led earth scientists in the last three centuries to establish the concept of an ancient and continuously evolving earth. Initially, the development of a geologic time scale was purely relative; geologists were only able to determine that certain isolated events took place before or after other isolated events. It was only much later—in the last few decades—that the absolute time scale of events could be measured in years.

Relative time

Late in the seventeenth century, Nicholas Steno, a Danish scientist, made two important observations. While apparently obvious and straightforward, they are fundamental to the dating of geologic events. His first observation was that sediments deposited by water accumulate in beds parallel to the surface of deposition. Because such surfaces are nearly always horizontal, the beds of sediment when first deposited are also horizontal. This is Steno's *Law of Original Horizontality*. The second observation was that in a pile of sediments, the older beds lie below the younger beds. This was his *Law of Superposition*. Both laws, of course, apply to original conditions of sedimentation. Steno recognized that after sediments are deposited horizontally, with the oldest below the youngest, they can be folded, uplifted, faulted, or overturned so that their original horizontality and relative age are not readily apparent.

It was not long after Steno that scientists interpreted individual sedimentary rock strata as the result of geologic processes like erosion, transportation, and deposition of sediment. And when earth scientists realized that sedimentary strata piled on top of each other or were contorted into odd configurations, they concluded that sedimentary rocks considered together reveal a history of geologic events. For instance, angular unconformities exposed in sedimentary rocks at Siccar Point, Scotland, led the great Scottish geologist James Hutton in 1788 to infer "the ruins of an earlier world . . . one land mass is worn down while the waste products provide the materials for a new one." The unconformity, seen in Figure 6–1, provided evidence of a sequence of geologic events including the

Undeformed, layered rock strata in Oregon illustrating the Law of Original Horizontality and the Law of Superposition.

deposition and lithification of the Silurian strata, their folding and erosion, and subsequent deposition of the Devonian strata on them. Although the absolute ages of these events were not known at the time, geologists correctly concluded that the events represented extremely long intervals of time. This was one early attempt by an earth scientist to establish the temporal ordering of geologic events by interpreting the rock record.

In the early nineteenth century, another British geologist, William Smith, observed that certain kinds of fossils were closely associated with particular rock strata. Moreover, the succession of these fossils, from oldest to youngest, was always in the same order. By studying a region's rocks and fossils, earth scientists began to identify the age of a stratum, even if it had been disturbed by faulting or folding after its initial formation. Before long, relative ages of more and more areas in Britain and Europe were established.

This has been called the Law of Floral and Faunal Succession. Why should this law hold true?

A homely analogy illustrates this principle. Suppose that you and I each collect our family photographs and arrange them temporally. Even if there were no dates on the pictures, we could put them in order by the approximate ages of relatives, clothing styles, car models, and other incidental information. For example, the picture of my late father with all his hair, wearing his college sweater, and standing proudly in front of his new rumble-seat coupe was obviously taken before the photograph of him with

Figure 6–1

An angular unconformity separating vertical strata of Silurian age and overlying tilted strata of Devonian age at Siccar Point, Scotland (see Table 6–1).

What age are the rocks of your area?

less hair, wearing his World War II civil defense armband. After each person's collection has been placed in time sequence, we might try to determine whose father was born first, who has the oldest picture, or what our own relative ages are.

In the same way, geologists can order rock strata by the "styles" of fossils they find. Primitive arthropods called trilobites come before a dinosaur, just as a rumble-seat coupe precedes a spaceship.

Using fossil sequences and the superposition of sedimentary rocks and taking into account other local information about faults, folding, and igneous rock contacts, earth scientists can reconstruct the geologic history of a local area, then a broad region, and eventually a whole continent. See Figure 6–2 for an example of this relative dating. In this way, scientists established a relative geologic time scale, as found in Table 6–1. Subdivision names in the scale refer to fossils (e.g., Paleozoic, "ancient life") or to geographic regions where rocks of that age were first discovered (e.g., Jurassic, for the Jura Mountains of Switzerland). Although many refinements were later added, the geologic scale of relative time was established by the late eighteenth century.

146

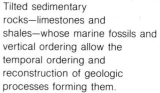

Tilted sedimentary rocks—limestones and shales—whose marine fossils and vertical ordering allow the temporal ordering and reconstruction of geologic processes forming them.

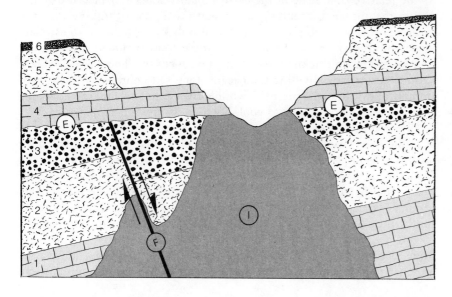

Figure 6-2

Hypothetical geologic cross section illustrating a sequence of events that formed rocks. Sedimentary rocks (1–3) were deposited, lithified, and tilted. The igneous intrusion (I) was emplaced either before or after tilting; faulting (F) of the intruded rocks occurred next. A period of uplift and erosion (E) then followed. Next, sedimentary rocks (4–6) were deposited, then tilted and eroded.

Table 6–1   Relative Geologic Time Scale

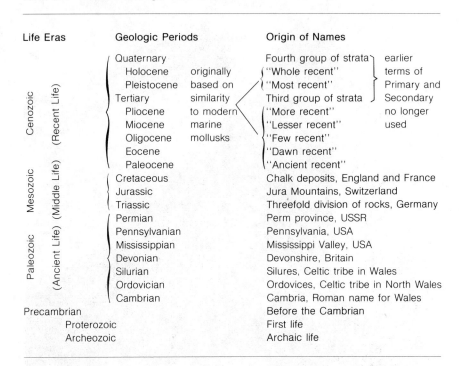

Life Eras		Geologic Periods		Origin of Names	
Cenozoic (Recent Life)		Quaternary		Fourth group of strata	earlier terms of Primary and Secondary no longer used
		Holocene	originally based on similarity to modern marine mollusks	"Whole recent"	
		Pleistocene		"Most recent"	
		Tertiary		Third group of strata	
		Pliocene		"More recent"	
		Miocene		"Lesser recent"	
		Oligocene		"Few recent"	
		Eocene		"Dawn recent"	
		Paleocene		"Ancient recent"	
Mesozoic (Middle Life)		Cretaceous		Chalk deposits, England and France	
		Jurassic		Jura Mountains, Switzerland	
		Triassic		Threefold division of rocks, Germany	
Paleozoic (Ancient Life)		Permian		Perm province, USSR	
		Pennsylvanian		Pennsylvania, USA	
		Mississippian		Mississippi Valley, USA	
		Devonian		Devonshire, Britain	
		Silurian		Silures, Celtic tribe in Wales	
		Ordovician		Ordovices, Celtic tribe in North Wales	
		Cambrian		Cambria, Roman name for Wales	
Precambrian				Before the Cambrian	
		Proterozoic		First life	
		Archeozoic		Archaic life	

Absolute time

As the relative time scale of the earth's history was being pieced together during the nineteenth century, some earth scientists, thinking that the earth was surely much older than several millenia, attempted to estimate its absolute age in years. Efforts along these lines included dividing the maximum known thickness of sedimentary rocks of all ages by an average sedimentation rate; dividing the oceans' total salt content by the annual amount of salt brought from the world's rivers; and calculating how long a molten earth would take to cool down to its present temperature.

Why would the cooling-of-a-molten-earth calculation give a result that is much too small?

All these estimates, which ranged from a few million to several tens of millions of years, were way off the mark because they were based on incorrect assumptions. Nevertheless, as approximate and inaccurate as they were, they were closer to reality than previous estimates of a few thousand years. Finally, when radioactivity was discovered around the turn of this century, earth scientists could determine the earth's total age. More recently, they have assigned absolute ages to parts of the relative time scale.

How are these absolute ages established by radioactivity? As you recall from Section 1–1, a chemical element is determined by the number of protons in its nucleus. Besides protons, the nucleus might contain one or more neutrons that have little influence on an element's chemical behavior but can modify its physical properties. Atoms of the same element with different numbers of neutrons are called *isotopes*. Oxygen, for example, has three isotopes: oxygen-16 (O^{16}), containing 8 protons and 8 neutrons; oxygen-17 (O^{17}), with 8 protons and 9 neutrons; and oxygen-18 (O^{18}), with 8 protons and 10 neutrons.

Some isotopes have such large nuclei, with numerous protons and neutrons, that they are unstable. They spontaneously break down, or decay, into stabler isotopes with smaller nuclei. Unstable isotopes are thus called *radioactive isotopes* because of their radiation activity; they radiate nuclear particles during atomic decay. Radiation energy is transformed to heat and generates much of the heat found within the crust and mantle.

Each radioactive isotope has its own rate of nuclear disintegration which is inversely related to its half-life. For example, the half-life of carbon-14 is about 5600 years, while that of potassium-40 is more than 1 billion years. Hence, the rate of radioactive decay in carbon-14 is far faster than that of potassium-40. Recall from Chapter 1 that the time it takes a radioactive isotope to decay or disintegrate to one-half its original mass is its half-life; see Figure 6–3. Varying physical and chemical conditions in the earth do not affect the half-life characteristic for each unstable isotope; this is strictly a time-related phenomenon. Some radioactive isotopes have short half-lives lasting just a few seconds or minutes. Others have half-lives measured in thousands, millions, or even billions of years, as indicated in Figure 6–4. Notice in Figure 6–4 that an unstable parent isotope decays into stable daughter isotopes that may be completely different chemical elements. Thus, U^{238} decays into Pb^{206} and helium, while K^{40} decays into A^{40} and Ca^{40}.

Shortly after the discovery of radioactivity, earth scientists began to use longer half-lived isotopes to date rocks. By measuring relative proportions of unstable parent isotopes to stable daughter isotopes, they could calculate absolute ages of rocks containing the isotopes. For example, the time when a granite was intruded can be determined by calculating the

Figure 6–3

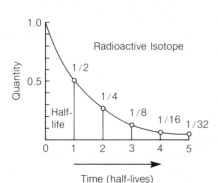

Decay of radioactive isotopes measured in half-lives. This graph illustrates the concept of half-life, the time it takes for a radioactive isotope to decay to one-half its original amount. After five half-lives have elapsed, only 1/32 of the original amount of the isotope remains.

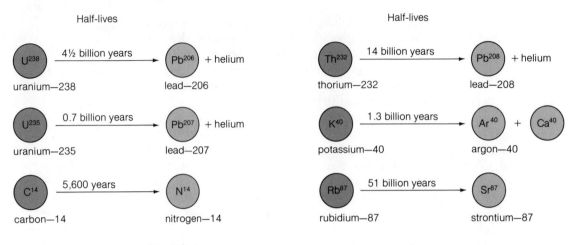

Half-lives

U²³⁸	→ 4½ billion years	Pb²⁰⁶	+ helium
uranium—238		lead—206	
U²³⁵	→ 0.7 billion years	Pb²⁰⁷	+ helium
uranium—235		lead—207	
C¹⁴	→ 5,600 years	N¹⁴	
carbon—14		nitrogen—14	

Half-lives

Th²³²	→ 14 billion years	Pb²⁰⁸	+ helium
thorium—232		lead—208	
K⁴⁰	→ 1.3 billion years	Ar⁴⁰	+ Ca⁴⁰
potassium—40		argon—40	
Rb⁸⁷	→ 51 billion years	Sr⁸⁷	
rubidium—87		strontium—87	

Figure 6–4

Some common radioactive parent isotopes and their decay daughter products used for absolute age-dating of rocks. The abundance ratio of each pair is proportional to the time of formation of the rocks containing them.

isotopic ratio of the unstable parent, potassium-40, found in the granite's potassium-bearing minerals like feldspar and mica, to its stable daughter, argon-40. The older the granite, the smaller the ratio of potassium-40 to argon-40. Although the theory behind radioactive age dating appears simple, in practice many difficulties must be overcome before reliable rock ages can be determined.

Once again, to grasp the late appearance of life—as well as humans—plot the absolute time scale on an 24-hour basis. Where would the building of the Egyptian pyramids, some four thousand years ago, appear on such a scale?

In the last few decades, geologists have established absolute ages for much of the relative geologic time scale by measuring isotopic ratios within rocks from all geologic epochs; their results are shown in Figure 6–5. By assigning absolute ages to the time scale, we can calculate with reasonable accuracy the rates of geologic processes recorded by rocks or sediments. For instance, a core of sediment sampled from the sea floor may indicate gradual, deep sea sedimentation. If we can determine absolute ages of the core's top and bottom, we then know how long it took for the sediment to deposit. If we divide the core's age by its length, we can determine the average sedimentation rate for that part of the ocean.

6–2 The Evolving Landscape

Landscapes are transformed by periodic uplift and erosion. In mountainous areas, landscapes stand thousands of meters above sea level and display jagged topographic relief. High elevations and pronounced relief indicate

150

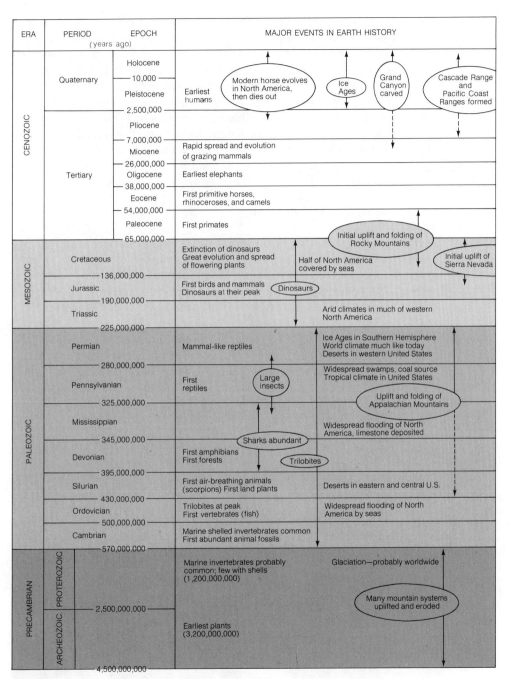

ERA	PERIOD	EPOCH (years ago)	MAJOR EVENTS IN EARTH HISTORY

Figure 6-5

Absolute geologic time scale showing important events in the
history of the earth and of life.

relatively recent uplift. Areas which were uplifted further back in time, and were therefore vulnerable to erosion for a longer time, stand lower and have less relief. Rates of landscape lowering by surficial processes decrease as elevation diminishes and topographic relief is more gentle.

Can you explain why these rates increase with topographic elevation and relief?

As we discussed in Section 3–5, once land masses erode, isostatic rebound raises them again, although not as high as they were during the previous erosion cycle. You will find an illustration of this in Figure 6–6. Complete erosion down to sea level does not occur because occasional mountain-building rejuvenates the landscape and the whole process of leveling begins again. The presence of unconformities and thick sequences of sedimentary rock proves that cycles of uplift and erosion never cease.

Rates of uplift and erosion

Land uplift is of two types: orogenic, or mountain building, and *epeiro-genic*, or broad regional upwarping. When continents converge at plate boundaries, orogenic uplift shapes mountain systems like the Andes and Himalayas. Epeirogenic uplift along continental margins and interiors has several causes, including distant orogenies and isostatic rebound after rocks have eroded or glaciers melted; see Section 3–5.

Although precise data for orogenic uplift rates are scarce and often contradictory, most earth scientists accept as an average maximum rate about 10 meters per thousand years. In Table 6–2 you'll see that rates of orogenic uplift for parts of California—an obviously tectonically active area—average 750 centimeters per thousand years. Maximum rates of uplift in the Soviet Union have been measured at 1000 centimeters per thousand years.

Figure 6–6

Erosional stages (a) Swiss Alps that are geologically young with high altitude and strong relief. (b) Australian Alps with moderate altitude and gentle relief. (c) Florida everglades lying near sea level and having low relief. (d) Rejuvenated landscape in Utah, where regional uplift has increased altitude and relief, permitting the meandering Colorado River to cut down deeply.

(a)

(b)

(c)

(d)

Table 6–2   Selected Tectonic Rates

Tectonic Processes	Some Typical Rates (centimeters per thousand years)
Mountain uplift (California)	
Santa Monica Mountains	390
San Gabriel Mountains	600
San Bernardino Mountains	1000
Post-glacial rebound	
Scandinavia	1100
Canada	1400
Epeirogeny	
coastal margins (average)	96
Faulting	
San Andreas, California	1300
Seafloor spreading	
North Atlantic	2500
East Pacific	8500

In regions where the last continental ice sheets melted quickly, isostatic rebound is relatively swift, and epeirogenic uplift rates approach those for orogenic uplift. We can assume that isostatic rebound in areas losing large rock masses by erosion would be more gradual because rock removal by erosion is surely slower than ice melting. Data on rising coasts around the world where epeirogeny is happening indicate that uplift rates are one-tenth those of orogenic uplift, or about 1 meter per thousand years.

When the land uplifts into mountains, it rises faster than it erodes, which is why mountains have high elevations and steep slopes. Yet erosion in such areas generally proceeds faster than in places with lower elevation and relief. Physical disintegration is greater at higher elevations because temperatures vary more widely—daily as well as seasonally—making the climate more severe and the vegetative cover sparser (Figure 6–7). Moreover, the steep slopes are less stable; materials are pulled downhill more easily by gravity. Apparently chemical alteration proceeds at rates independent of elevation. Such weathering is determined more by climate and type of bedrock than by altitude.

As the land becomes more level, erosion rates fall off accordingly. For instance, average erosion rates for the Colorado River, draining parts of the Rocky Mountains and high Colorado Plateau, surpass those for the less elevated slopes of the Pacific coast. Erosion rates for the Mississippi River and the Atlantic coast rivers, which drain regions of still lower elevation and gentle topographic relief, are about one-third those for the Colorado River—as indicated in Table 6–3. Erosion rates for the world's highest mountain system, the Himalayas, have been estimated as high as 100 cm per thousand years—more than sixteen times the average U.S. rate of erosion.

Perhaps you have noticed an interesting paradox: if the uplift rate is at least ten times faster than erosion rate (1 meter per thousand years for slow uplift, compared to 10 centimeters per thousand years for rapid erosion), why doesn't the land keep moving ever upward? The apparent discrepancy is explained by the fact that land erosion is continuous, even if variable from place to place, whereas uplift—whether epeirogenic or orogenic—is discontinuous and episodic. The opposing processes are illustrated in Figure 6-8. Hundreds or thousands of years may pass with no uplift, during which time erosion is constantly wearing away the land. Perhaps every few hundred thousand years epeirogeny occurs, or every few hundred million years orogeny, thereby increasing the land's elevation as well as the erosion rates. Then there may ensue a long interval of erosion with little or no uplift.

Calculate how long it would take to erode some of the world's highest mountains down to sea level.

Collecting sediments

What happens to the materials eroded from the lands? As we discussed in Section 3-6, materials eroded from land are deposited along continental shelves as thick wedges of sediment which taper off on continental rises. For example, erosion of the continental interior of the United States results in the deposition of a thick mantle of sediments around its seaward periphery. Table 6-3 shows sedimentation rates for the Colorado, Mississippi, West Coast, and East Coast rivers of the United States. Differences among drainage areas are due to variations in erosion rates and sizes of the drainage basins. Although the Colorado River erodes more than three times faster than the Mississippi, the drainage area of the Mississippi is more than five times larger than that of the Colorado. Consequently, the Mississippi River deposits 50 percent more sediment at its mouth.

A metric ton contains 1000 kilograms and is equal to 2205 pounds. Thus, 1 kilogram is equivalent to 2.2 pounds.

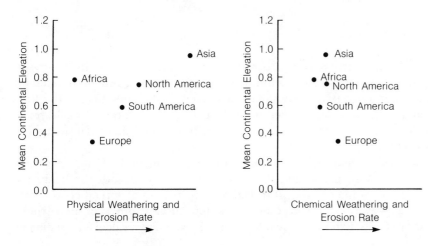

Figure 6-7

Graphs showing that the rate of physical weathering and erosion increases with continental elevation, whereas the rate of chemical weathering and erosion is independent of continental elevation. For reasons that are not clear, Africa has lower rates for both kinds of weathering and erosion.

Table 6–3 Selected Rates of Erosion and Sedimentation

Earth Processes	Rates
Erosion	(centimeters per thousand years)
Colorado River drainage area	17
Pacific slope of California	9
Mississippi River	5
North Atlantic drainage area	2
average U.S. rate	6
Sedimentation[a]	(million metric tons per year)
Colorado River	281
Pacific slope	76
Mississippi River	431
North Atlantic Coast of U.S.	48

[a]Includes dissolved and suspended load.

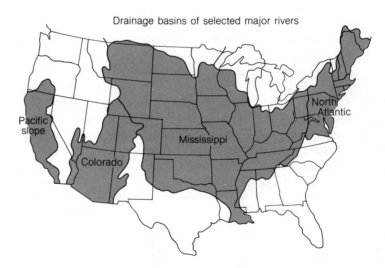

Drainage basins of selected major rivers

In deeper, offshore parts of oceans, sediments slowly collect in the abyssal plains of ocean basins. Accumulation rates average a few millimeters to a few centimeters per thousand years. These sediments include the fine-grained silts and clays eroded from the continents and the tiny calcareous (calcium-containing) and siliceous (silica-containing) shells of dead marine plankton.

Remember that some plates travel toward continents and may have their deep-sea sediments skimmed off as they descend beneath the landmass, piling the sediments against it. Plates move at rates of several centimeters per year—away from spreading ridges, toward each other, or along trans-current faults. Refer to Table 6–2 for sample rates. From calculating sediment thicknesses along the United States' Pacific and Atlantic Coasts we know that the Atlantic Coast has about five times as much sediment as the Pacific Coast, even though the Pacific drainage basins are somewhat larger and erosion rates higher. This discrepancy comes about because the Pacific Coast is on the leading edge of the American plate while the Atlantic Coast is on the trailing side. Hence sediments on the Pacific side are swept against and under the American plate, whereas those on the Atlantic side ride with the plate and accumulate. The effects of differential accumulation along the continental margins of the east and west coasts are twofold. First, the continental shelves are much wider on the east than on the west. Second, sediments on the west are deformed as they pile against the continent or are brought down along the subduction zone; those on the eastern side are flat-lying and dip gently seaward.

6-3 The Story behind the Rates

Rates of uplift, erosion, and sedimentation are often calculated in such a way as to average seasonal or annual fluctuations in rates. Earth scientists average the fluctuations so they can obtain reliable estimates with long-term, geologic significance; they want figures that describe processes operating over long intervals of geologic time, measured in thousands of years or more. However, short-term variations in these rates are also of interest, particularly for earth scientists concerned about environmental

Wolman's Viewpoint at the end of this chapter emphasizes the need for such specific data regarding human impact on the earth.

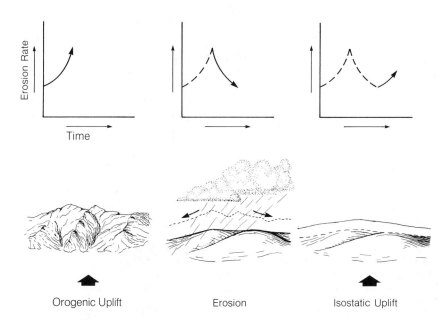

Figure 6–8

Hypothetical graphs of uplift and erosion. Episodic orogenic and isostatic uplift results in increases in erosion rates, which decrease until the next uplift.

questions. A geologic engineer, for example, is much more interested in knowing how frequently a river may overflow its banks in the next decade than what its erosion rate has averaged over centuries.

Variations in geographic scale are also important considerations in discussing geologic rates. The larger the geographic area for which a rate is calculated, the less the rate is likely to vary. For instance, the erosion rate of the Mississippi River at its mouth on the Gulf of Mexico will vary less, proportionally, than the erosion rate within one of its smaller tributaries far upstream; the latter is much more susceptible to local variations in rainfall, mass wasting, human interference, and so on.

Frequency and magnitude

In addition to differences in time and geography, there is another pair of variables affecting the rate of geologic processes. We might assume that a geologic process which operates rarely, but at a large magnitude, has greater impact than one which operates more frequently but at smaller magnitudes. For instance, we could expect more sediment to be transported by a rare flood of great magnitude than is carried by a river in its more usual state. Although it is true that the amount of work done (erosion, sedimentation, and so on) increases exponentially with the force of a

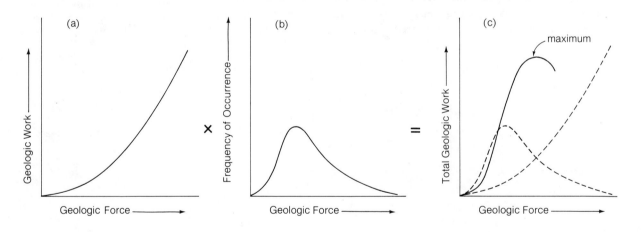

Figure 6–9

Graphs of the work done by geologic forces. (a) The amount of work done increases at an accelerating rate as the force applied increases. (b) The frequency of given amounts of force vary, with few small or large forces and many of intermediate range. (c) The total work done by the force is a product of the magnitude of force times its frequency. Note that maximum work is done by moderate forces of intermediate frequency.

particular event, we must also keep in mind how frequent such events are. The net work accomplished by a geologic process (mass wasting, stream flow, surf action, and so on) is the product of the magnitude of each event times its frequency, as explained in Figure 6–9. Most geologic work is accomplished by moderate-sized events of medium frequency.

In rivers, for example, most suspended sediment is transported downstream by water flows of moderate magnitude, as you will find in Figure 6–10. A moderate-magnitude flow is somewhat greater than the average daily flow, and it occurs only a few times a year. Moreover, when a river's flow is variable, more sediment load is likely to be carried by larger, infrequent flows. Thus the rarer but large-magnitude, once-every-50- or- 100-year flood may do *less* net geologic work than the smaller, more frequent annual floods. The latter, in turn, do *more* work than the everyday flow when the river is not in flood. In short, the occasional flood of moderate magnitude does more work than either the rare, big flood or the daily, small regular flow.

Beach erosion exhibits similar traits. The slope of a beach depends on the grain size of its sediment which, in turn, is controlled by the steepness of waves (ratio of wave height to wave length) attacking it. The particles in coarse-grained sediments tend to be angular and rough-surfaced, while

Look at Figure 6–10 again; in how much time do the rivers carry 90 percent of their suspended load?

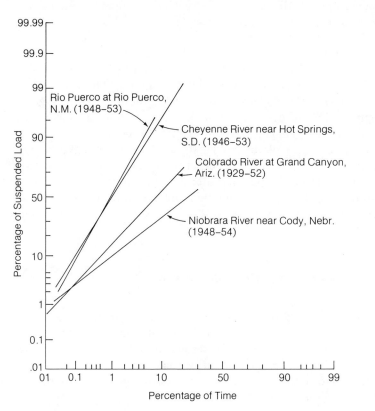

Figure 6–10

Graphs of four rivers showing the time required to transport various percentages of their total suspended load. In general, most of the suspended sediments of these rivers is carried during only a short part of the year. The Rio Puerco carries most of its load during short, intense summer and fall cloudbursts. Similarly, the Cheyenne River experiences large variations in flow and transports most of its material during summer rains. The Colorado and Niobrara Rivers have a more uniform flow, but even they carry most of their sediment during infrequent periods of high flow. Note that 50 percent of the river's flow is carried only about 4 days of the year for the Rio Puerco and Cheyenne, 31 days for the Colorado, and 95 days for the Niobrara.

159

the particles in fine-grained sediments are smoother and more rounded. Consequently, coarser materials usually can maintain slopes a few degrees steeper than fine-grained ones. Pebbly beaches thus have steeper slopes than sandy beaches; and if grain size is constant, beaches facing relatively steep waves will have higher slopes than beaches facing waves with smaller height-to-length ratios.

Wave steepness is determined by wind velocity and *fetch*, or distance across which the wind blows. Thus high winds blowing across long distances form steeper waves than light winds blowing over short distances. Now when beach slopes are measured, the slopes are found to be in equilibrium with the average local wind velocity (with some variation from winter to summer). Winter beaches are steeper because they are attacked by short, choppy waves stirred up by stronger and more frequent winter storms. Summer beaches are less steep because summer waves roll in as low, long swells, and summer winds are less strong. The occasional big storm, winter or summer, greatly disturbs a beach profile, but prestorm conditions usually return in a few weeks or months. In other words—and this is the point—the shape of a beach is determined by average summer or winter conditions, and not ordinarily by infrequent big storms.

So for rivers and beaches, we can measure both short-term and long-term fluctuations in the processes which affect them. What about processes that do not operate continuously, but instead are irregular or episodic? A different kind of activity rate can be estimated by collecting data on the frequency and magnitude of discrete events like volcanic eruptions, landslides, or earthquakes. For example, historical records of earthquake activity in a given area can be plotted graphically to show how frequently earthquakes of a certain magnitude hit the area. From a graph such as Figure 6–11, we can predict with some confidence that an earthquake of a particular magnitude will occur. We could estimate, for instance, that for the area plotted in Figure 6–11 an earthquake of magnitude 6 or greater will probably happen at least once during the next 30 years—which might be the planned lifetime of a proposed man-made structure. Admittedly, these predictions are more like educated guesses than absolute certainties, but they do give some sense of how often episodic events might strike.

6-4 Travels in a Closed System

The earth is a closed system as well as a dynamic one. Not only does it continually experience change, but materials within the system periodically move throughout it. Therefore, in addition to calculating how fast certain earth processes proceed, we can estimate also how long it takes materials to complete a cycle within a particular earth system. This interval is called *residence time*.

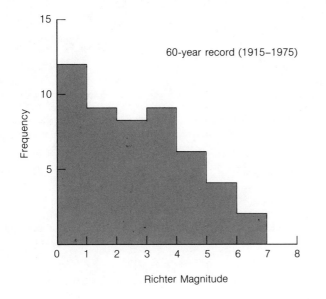

60-year record (1915–1975)

Figure 6–11

Example of a frequency-magnitude graph for episodic geologic events. Shown here is a hypothetical graph of the frequency of various magnitude earthquakes that occurred in an area over a 60-year period. There were many small magnitude earthquakes and several large ones.

Earth materials complete their cycles at varying speeds. As noted in Chapter 1, oxygen recycles in the atmosphere-hydrosphere-biosphere in 2000 years, whereas carbon dioxide takes only 300 years. Oxygen therefore has a longer residence time than carbon dioxide. The reasons behind these different rates are the variations in speed among earth cycles and the relative volume of material being recycled. With this knowledge we can calculate the residence times for different materials. For example, suppose we know the volume of dissolved substances in the oceans and the annual rate at which they enter the seas from the world's rivers. To estimate the residence time of these substances in the oceans, we divide their volume by their rate of entry. This calculation works if we assume that the oceans' salinity does not change, because the amount of salts added to the oceans equals the amount removed by sediments or organisms and returned to continents by moving oceanic plates. That is, the average amount of dissolved salts added annually is equal to the average amount removed annually. Although such calculations require some assumptions and are therefore imprecise, they do give us a generally reliable estimate of how long materials remain within a geologic system before being re-cycled. Table 6–4 summarizes residence times for important materials circulating in the atmosphere, hydrosphere, and biosphere.

Refer back to Figure 1–7 where several cycles are illustrated.

Circulation within the atmosphere

The earth's atmosphere consists of several vertical layers having charac-teristic temperatures, pressures, and densities. Within the *troposphere*, or

Table 6–4   Selected Residence Times for Earth Materials

Earth Materials	Some Typical Residence Times
Atmosphere circulation	
water vapor	10 days (lower atmosphere)
carbon dioxide	5 to 10 days (with sea)
aerosol particles	
stratosphere (upper atmosphere)	several months to several years
troposphere (lower atmosphere)	one to several weeks
Hydrosphere circulation	
Atlantic surface water	10 years
Atlantic deep water	600 years
Pacific surface water	25 years
Pacific deep water	1300 years
terrestrial groundwater	150 years [above 760 meters (2500 feet) depth]
Biosphere circulation[a]	
water	2,000,000 years
oxygen	2000 years
carbon dioxide	300 years
Seawater constituents[a]	
water	44,000 years
all salts	22,000,000 years
calcium ion	1,200,000 years
sulfate ion	11,000,000 years
sodium ion	260,000,000 years
chloride ion	infinite

[a]Average time it takes for these materials to recycle with the atmosphere and hydrosphere.

lower atmosphere where most recycling occurs, water vapor recycles rather quickly. Water evaporated from land and sea surfaces accumulates in the troposphere and falls again as snow or rain within about ten days. Carbon dioxide, generated by oxidation of organic matter, animal and plant respiration, and combustion of fossil fuels like coal, oil, and gas, also has a short residence time. About one-third of the carbon dioxide introduced in the atmosphere mixes with the oceans and goes into solution with sea water. Very fine particles of dust and ash from the land and salt from the sea—so-called *aerosol particles* (*aer*, Greek for air, and *solidus*, Latin for solid)—mix with both the lower and upper atmosphere, or *stratosphere*. These aerosols are carried back to the earth's surface by snow and rain—within several weeks in the troposphere and within several months to several years in the stratosphere.

Although circulation in the atmosphere varies according to season, latitude, and the type of materials circulating, residence times here are quite short compared with those for materials in the hydrosphere and biosphere. If you look again at Table 6–4, for example, you will discover large variations in the time it takes water, oxygen, and carbon dioxide to travel through the atmosphere and hydrosphere and back again to the biosphere. The differences are mainly due to corresponding variations in the volume of water, oxygen, and carbon dioxide available on the earth's surface.

We say more about these variations in the next section.

Volume's role in recycling

The volume or size of a resource like water, oxygen, or carbon dioxide raises an additional interesting question about recycling and residence time. If a resource is used at a constant rate, the time it takes to recycle that resource is proportional to the volume of the resource: a larger volume will recycle more slowly than a smaller one, and vice versa. Additionally, if two different resources have the same volume, the relative recycling speed for each will depend on how fast the resource is used. To cite an example, the residence time of nitrogen in a grassland is short compared with the residence time of carbon because nitrogen is typically scarcer than carbon. But for a *given level* of nitrogen, its residence time in a grassland is shorter than in a forest. The grasses recycle annually; the seeds germinate and grow in the wet season with the grasses making new seeds and dying in the dry season. Thus nitrogen bound up in one year's crop is released by decay organisms for next year's crop. In a forest, though, nitrogen is held for a greater interval by the longer-lived trees. Annual plant growth depends on nutrients released when leaves or some trees die and decompose; each year's decomposition releases only a fraction of the total nitrogen bound up in the living forest. Hence, the residence time of nitrogen specifically, but any substance in general, is determined both by its volume and its rate of use.

This principle applies well to the residence times of water in the oceans and on the land. The much smaller volume of water that lies at shallow depths within the ground recycles about every 150 years. This includes the time it takes to fall from the sky, percolate down into shallow groundwater, and move slowly through springs to streams and rivers where it is evaporated. Shallow depth water in the Atlantic and Pacific Oceans recycles faster than every 150 years because these surface waters are actively circulated by waves and currents; thus the water evaporates and mixes with the atmosphere more rapidly. Deep ocean water, however, circulates more slowly and as a consequence has longer residence time. Variations in residence times between the Atlantic and Pacific result from their different volumes.

Returning again to Table 6–4, you'll notice that salts brought to the sea vary greatly in their respective residence times. The abundant and highly soluble sodium and chlorine salts recycle more slowly than the less abundant and less soluble sulfate and calcium salts. Despite these variations, the large volume of all oceanic salts results in long residence times.

6–5 Variations among Cycles

By now, concepts of change within the earth over time should be clear. Moreover, you should have some notion of how to calculate rates of change and residence times on broad regional or even global scales for fairly long periods of time. Yet it may have occurred to you that it would be difficult to apply such calculations to more local situations over shorter intervals of time. For example, what about the following questions? What is the residence time of toxic chemicals dumped off the coast of New York City? Will radioactive materials placed in deep wells decay to harmless levels before they appear in surface or shallow groundwater? How long before sewage dumped far offshore in deep water reappears along local beaches? Will it have had sufficient time to decay and dilute naturally? How many earthquakes with a magnitude of 6.5 or more will a nuclear power plant or large hydroelectric dam experience during their projected lifetimes?

Each of these questions can be answered only in the local geologic context for which they are posed. But, realizing that *all* earth systems involve change and recycling, earth scientists are now attempting to determine local rates and residence times relevant to many specific environmental issues.

Thresholds, triggers, and feedbacks

Other factors that we'll refer to as thresholds, triggers, and feedbacks complicate estimates of rates and residence times. Besides differences in temporal and geographic scale, we can find variations in rates of geologic activity that, instead of being gradual, take big jumps up or down in rate. Ordinarily, we might assume that as we turn up the rate of some geologic process, the results will be proportional to the rate increase. But this is not always so; mechanisms such as *thresholds, triggers,* and *feedbacks* may also come into play.

Let's first consider threshold phenomena operating within natural systems. Suppose that we add small amounts of organic matter to natural surface waters. Oxygen dissolved in the surface water oxidizes the organic matter and produces mostly carbon dioxide gas and water. Oxygen-using animals and plants in the water, of course, compete with the organic matter, but up to a certain point we notice no change in the system as we keep dumping the organic waste. However, when we reach that certain point, or threshold, when organic matter is so abundant that it prevents organisms from getting enough oxygen, the whole system rapidly becomes oxygenless or stagnant with dramatic results. Animals and plants die, contributing still more organic matter, and the excess organic wastes that are not oxidized begin to accumulate. Chemical reactions between the organic matter and the bacteria that do not use oxygen for respiration generate hydrogen sulfide gas. This gas is toxic to other organisms and further deteriorates the natural waters. Before long the water body is lifeless, smells like rotten eggs, and, most seriously, is unable to recover its original condition.

Another factor that causes sharp changes in rates and residence times is when a phenomenon triggers such a significant change in a geologic process that it operates faster than normal. By burning local vegetative cover, a forest fire may trigger accelerated erosion and mass wasting far beyond the usual rate. During the summer of 1972, in Big Sur, California, a forest fire burned vegetation covering over 1700 hectares. The following fall, heavy rains brought widespread landslides and mudflows. Roads and buildings were destroyed by tremendous quantities of soil and rock moving rapidly downslope. This area had received similar heavy rains before without such extensive mass wasting and erosion, but this time the forest fire triggered much faster rates by removing the protective vegetative cover.

One hectare is equal to 10,000 square meters and is equivalent to 2.5 acres.

A third and final consideration in our discussion of rates and cycles is feedback mechanisms. Feedback refers to a situation in which a cause produces an effect which, in turn, influences the cause. Feedback may be negative, in that the causal force is diminished and lessened in its effect. Or feedback may be positive, enhancing the causal force and producing a still larger effect.

A graded stream is a geologic example of a negative feedback system. For instance, an increase in water discharge heightens stream velocity; this in turn erodes a wider and deeper channel and lowers the gradient. The gradient decreases until the extra energy of the new discharge is absorbed in the work of transporting the higher sediment load over the lowered slope. If, on the other hand, the sediment load increases without a proportionate increase in discharge, the river will make its gradient steeper (as in a braided river) by depositing sediment until the steeper gradient and consequently greater water velocity can accommodate the larger load. Thus, rivers can adjust to changes in discharge and sediment load by negative feedback mechanisms. Most geologic processes are regulated by such negative feedback.

The concept of dynamic equilibrium discussed in Cloud's Viewpoint in Chapter 1 is essentially a negative feedback phenomenon.

Examples of positive feedback are those situations we refer to as "vicious cycles"; that is, the effect of a cause feeds back to the cause and creates still larger effects. A common sociological example of positive feedback is the effect of poverty in diminishing educational opportunity, which in turn makes people less employable and thereby worsens their poverty. A geologic example of positive feedback is the upset of a natural, steady-state system such as sediment transport along a beach by longshore currents. In its natural state, the beach is in equilibrium: sediment transported down the beach by longshore currents is replaced by sediment from up the beach. Although sediment is constantly moving, the beach remains more or less the same. Attempts to control or stop sediment transport by building jetties or breakwaters usually result in greater sediment deposition, or erosion, within the system, thereby requiring still more artificial construction to control the situation.

Viewpoint   M. Gordon Wolman

M. Gordon Wolman is Professor of Geography, Johns Hopkins University, Baltimore. Dr. Wolman's special interest in his hydrological studies is the impact of a variety of human activities on natural waterways. In this Viewpoint Dr. Wolman emphasizes the need for more precise data about the ways in which natural systems are altered by man's works.

The Known and the Unknown

In 1969 a judge in California was required to decide whether or not lumbering should be permitted to continue in a forest of majestic redwoods. Opponents of the lumbering operation contended not only that erosion caused by the lumbering was wearing away the land but that the resulting deluge of mud was destroying the trout streams. The lumber company argued that the effects were not so devastating, and that in any case they would control the erosion if told what to do. Hearings before the judge took weeks. In making his final decision, the judge had this to say about the problem of sediment:

> While numerous expert witnesses in the field of geology, forestry, engineering, and biology were presented, their conclusions and the opinions they derived from them are hopelessly irreconcilable on such critical questions as how much and how far solid particles will be moved by any given flow of surface water. They were able to agree only that sediment will not be transported upstream.

Considering that this book and many like it contain pages describing our knowledge of water, sediment, and the landscape, it may appear strange that none of the experts could provide the judge with the precise answers he needed. From the judge's standpoint, since the question could not be answered with the necessary precision, not much is known, perhaps even "nothing useful." But clearly, as the text shows, we professors and the

You sure about that?

thousands of scientists and engineers in the field have learned a great deal about the subject. What vexed the judge is, of course, an exciting opportunity to the student. What, then, are a few of the unknowns, the unanswered questions lying behind the apparent knowns, or answers, presented in the text?

The notion of equilibrium or balance is used in the text to describe how the hydrologic budget works, and the way in which landforms and stream channels may be determined. Common observations suggest simple evidence of the equilibrium forms of rivers (Chapter 4). Everyone knows that big rivers carry much more water than small rivers, that big rivers have large bends and small rivers have small bends, and that the amount of water in a river varies from a trickle to a flood although the river may look much the same over a period of years. Further, if a truckload of gravel is dumped into the river, the flow will move some of the gravel downstream, will redistribute other pieces, and the gradient may steepen, allowing the gravel to be more readily transported.

The relation of form to flow and the response to change suggest that the concept of equilibrium may be a useful way of thinking about how rivers behave. The concept may even provide a basis for prediction. For example, if the landforms and river systems are in equilibrium, then it will not be a surprise that when farmland is converted to cities, the supply of mud and water from the drainage basin will be altered. We can also assume that if the supply of water and sediment changes, then stream channels adjusted to the original conditions will be modified by the new conditions. Simply put, it should come as no surprise that an alteration of the landscape will produce what current jargon calls "environmental impacts," perhaps distant in time and space from the site of the disturbance.

Clearly, even this amount of knowledge can affect public policy. Yet the judge, or others faced with decisions, must know not only that there will be environmental impacts, but where these impacts will appear, how large they will be, how long they will continue, and what values to ascribe to them. At this stage the details matter a great deal, and more knowledge becomes necessary.

For the sediment-water problem, then, what are a few of the unknowns in the equation relating water, sediment movement, and river systems? Let's start with the watershed. How does rain falling on the watershed get into a stream? As the text indicates, during a rainstorm some water flows over the ground and some is absorbed by plants and soil. Water may arrive at the stream channel after traveling within the ground, within the surface layers of the soil, or over the ground surface. The particular path that the water takes makes a difference. For example, water that travels through the ground is likely to accumulate higher concentrations of salts; water flowing over the surface erodes the soil; the retarding of surface flow may save water; the slowing of infiltration may waste water. Studies have shown that water rises almost instantaneously in some small streams with each burst of heavy rainfall. The time between the burst and the rise is too short for each drop to have moved underground from a long distance. Measurements suggest that this rise may result almost entirely from water that falls directly on the channel or on wet areas adjacent to the channel. Evidence from the laboratory indicates, however, that if the ground is saturated, a drop of water falling on the landscape may simply increase the pressure in the saturated soil and an equivalent drop will appear instantly in the stream channel, just as if the raindrop fell directly on the stream. It is difficult to determine which process actually takes place. As yet, we cannot predict precisely how water arrives at channels in different terrain, in different climates, and at different times.

Knowing that the quantity of water a stream will carry can be determined, the California judge wanted to establish how much sediment would be eroded from the watershed, what the source areas would be, and how the sediment would move, once in the channel. The quantity of water in the channel fluctuates with the rainfall, and larger particles move only in higher flows. Sand may travel in dunes or waves, and fine silt may cloud the water for days after a storm. The quantity of sediment that moves in a channel in relation to the quantity of water can vary a great deal. The amount of sediment in a stream depends upon the amount and intensity of rainfall, the season, vegetation, topography, geology, and land use. Because sediment, along with the organic and inorganic materials that attach themselves to it, is one of the greatest "pollutants," we would like to know the sources of its supply and the processes that determine this supply. Many pollutants, such as trace metals and pesticides, cling to fine sediment particles. In our efforts to clean up the rivers, we would like to be able to estimate how long it will take pollutants already in

168

the rivers to be removed, or altered, and where they will ultimately be deposited. Such knowledge is essential in choosing effective, efficient, and economical policies to manage the environment.

Sediment moving in the flow or deposited on the bottom of a river may completely alter the habitat for fish, floating life, vegetation, and organisms dwelling on the river bottom. As long as the supply of material is too great for the stream to move, sediment will accumulate, blanket the bottom, fill up the pools, and filter between the rocks. Spawning sites in sands and gravels of trout and salmon streams may be decimated. Insects on which the fish feed may be destroyed, and deep clear pools filled with mud. Organisms that were once adapted to the stream will be displaced by different organisms adapted to the new conditions. When the supply of sediment or mud ceases, however, some streams may cleanse themselves, restoring the old habitats, while others may remain clogged. Thus, to estimate the damages produced by a given change in land use and to estimate the time of restoration, we need to know where the sediment comes from, what effect it has on organisms, and how long it takes to move through the stream system. We do know these things in a general way, but clearly not to the judge's satisfaction.

As Chapter 4 suggests, river channels "adjust" to the quantity of water and sediment delivered to them. What happens to the channel itself when land use changes and quantities of water and sediment change? As urbanization proceeds, we can predict a series of changes in the sediment yield. Farming and later urban construction may produce large amounts of sediment that clog the river channels. These deposits make the channel smaller. Because less water can be accommodated in the channel, flooding increases and, if trees grow on the deposits, the problem worsens. After urbanization is complete, no sediment comes from streets and roofs. As a result, frequent flows of runoff water erode and enlarge the stream channels without replacing the material with sediment from the watershed, as they did before the land-use change. We can predict that stream channels in city parks, for example, will erode and enlarge as more and more land on the watershed is urbanized.

What is missing? To do the job properly, we should be able to estimate not only that the channel will grow larger, but what the dimensions and characteristics of the resulting channel will be. We must ask the following questions: How wide and how shallow will the channel be? Will the bottom be coarse sand, cobbles, or debris? Will it continue to widen forever? We know that such widening has taken place in urban rivers. But park managers and planners would like to be able to predict how a stream in a park will develop. They would like to know when they should provide erosion control to protect the banks and assure that the stream will continue to flow and will not degenerate into a series of mosquito breeding puddles. Sometimes storage could be provided to control the runoff from city streets, but how much storage and at what cost?

Our habit, of course, is to throw money at such problems by building reservoirs, lining channels with concrete, or raising walls. But often such efforts carry with them low flows, aesthetic degradation, and only illusory safety. Then, too, we may sometimes throw money "down the drain" when the works we build don't do the job we expected. We should therefore try to determine not only what our dollars are buying, but also what alternatives might be available to achieve the same ends. Of course, even if no judge ever asked, it would be nice to know the odyssey of one grain of sand loosed high in the Rocky Mountains by a spring torrent and deposited in the Gulf of Mexico by the mighty Mississippi.

Summary

Earth processes operate continuously or episodically throughout time. The earth is thus a dynamic body constantly experiencing change, and the sequence of such changes can be ordered to establish a relative chronology of geologic events. Absolute dating of these events is possible through historical records and radioactive age-dating of the rock record. Establishing absolute dates for these events and changes enables us to calculate how fast the processes effecting change occur.

Earth processes also tend to be cyclical. Thus, earth materials within a natural system are circulated over intervals of time that depend on how fast the system operates and how much material is being circulated. We can calculate residence times when these rates and volumes are known.

Rates and residence times vary according to the size of the geographic area and the time interval under consideration. Most rates and residence times have been calculated for broad geographic regions over periods of hundreds to thousands of years or more. To understand temporal changes and recycling in geologic situations of more human scale (e.g., covering areas of a few thousand kilometers over one or two human generations), different rates and residence times must be calculated. Global averages may bear little relevance to local problems of environmental geology.

Many geologic processes operate as balanced negative feedback systems. Rare natural events or, more commonly, human intervention within these systems, may trigger rapid changes or introduce positive feedback that disturb these systems' natural equilibrium.

Glossary

feedback   Process in which a cause produces an effect that in turn influences the cause itself. Positive feedback increases the initial causal force whereas negative feedback diminishes it.

isotope   Chemical element having the same number of protons in its nucleus as another element, but a different number of neutrons. Isotopes of the same element thus have the same atomic number and almost identical chemical properties, but they differ in their atomic masses and physical behavior.

Law of Original Horizontality   Sediments are deposited in the earth's gravitational field and hence initially laid down horizontal to the earth's surface. Subsequent deformation may change this original horizontality.

Law of Superposition   As sediments are deposited, the older are laid down under the younger, or later, ones. Hence, in an undisturbed pile of sediments and sedimentary rocks, the oldest layers lie at the bottom, the youngest at the top.

radioactive isotopes   Isotopes that spontaneously break down into other isotopes or elements, yielding radiation as they do so.

residence time   The average amount of time a particular substance spends within a designated earth system. The residence time is inversely proportional to the rate of movement within the system and directly proportional to the size of the system.

threshold   A given level of some condition within an earth system up to which there is little apparent change and beyond which changes are rapid and of great magnitude.

trigger   The presence of some condition within an earth system or process that radically increases the rate at which that system or process operates.

Reading Further

Berry, W. B. N. 1968. *Growth of a Prehistoric Time Scale.* San Francisco: W. H. Freeman. Historical account of the three-century development of the relative geologic time scale.

Eicher, D. 1968. *Geologic Time.* Englewood Cliffs, N.J.: Prentice-Hall. A short introduction to the measurement of relative and absolute geologic time.

Wolman, G., and J. Miller. 1960. "Magnitude and Frequency of Forces in Geomorphic Processes." *Journal of Geology,* vol. 68, pp. 54–74. Analysis of the geologic work done by low-magnitude, frequent earth processes compared with that performed by the rarer, larger-magnitude phenomena.

Conversion Tables (Approximate)

English to Metric When You Know	Multiply by	To Find
inches	2.54	centimeters
feet	0.30	meters
yards	0.91	meters
miles	1.61	kilometers
square inches	6.45	square centimeters
square feet	0.09	square meters
square yards	0.84	square meters
acres	0.40	hectares
square miles	2.6	square kilometers
cubic inches	16.4	cubic centimeters
cubic feet	0.27	cubic meters
cubic yards	0.76	cubic meters
cubic miles	4.19	cubic kilometers
ounces	28.3	grams
pounds	0.45	kilograms
tons	0.9	tons
fluid ounces	30	milliliters
quarts	0.95	liters
gallons	3.8	liters

Metric to English When You Know	Multiply by	To Find
centimeters	0.39	inches
meters	3.28	feet
meters	1.09	yards
kilometers	0.62	miles
square centimeters	0.15	square inches
square meters	11	square feet
square meters	1.20	square yards
hectares	2.47	acres
square kilometers	0.38	square miles
cubic centimeters	0.06	cubic inches
cubic meters	0.37	cubic feet
cubic meters	0.13	cubic yards
cubic kilometers	0.24	cubic miles
grams	0.04	ounces
kilograms	2.20	pounds
tons	1.1	tons
milliliters	0.033	ounces
liters	1.06	quarts
liters	0.26	gallons

Energy

1 barrel of crude oil = 42 gallons

7 barrels of crude oil = 1 metric ton = 40 million BTUs

1 metric ton of coal = 28 million BTUs

1 gram U_{235} = 2.7 metric tons of coal = 13.7 barrels of crude oil

1 BTU (British Thermal Unit) = 252 calories = 0.0002931 kilowatt-hour

1 kilowatt-hour = 860,421 calories = 3412 BTU

Credits

Photographs

Page 1: National Park Service.
Chapter 1 Page 2: National Aeronautics Space Agency (NASA). Page 4: Lick Observatory. Page 12: Japan National Tourist Association. Page 16: Japan National Tourist Association. Page 19: High Altitude Observatory. **Chapter 2** Page 26: David Prowell. Page 37, left: National Park Service; right: National Park Service. Page 38: J. R. Stacy, U.S. Geological Survey. Page 44, top left: P. Jay Fleisher, © 1972, Harper & Row; bottom left: E. Bailey, U.S. Geological Survey; right: M. Woodbridge Williams, National Park Service. Page 45, top: J. R. Stacy, U.S. Geological Survey; bottom: P. Jay Fleisher, © 1972, Harper & Row. Page 46: U.S. Geological Survey. Page 49: P. Jay Fleisher, © 1972, Harper & Row. Page 51: William Sacco. **Chapter 3** Page 56: Naval Electronics Laboratory Center. Page 66, left: P. Jay Fleisher, © 1972, Harper & Row; right: U.S. Geological Survey. Page 72: National Park Service. Page 73, top: Swissair; bottom: Geological Survey of Canada, Ottawa. Page 74: P. Jay Fleisher, © 1972, Harper & Row. **Chapter 4** Page 84: KLM Aerocarto. Page 86: P. Jay Fleisher, © 1972, Harper & Row. Page 87: P. Jay Fleisher, © 1972, Harper & Row. Page 88, left: Swissair; right: Australian Information Service. Page 90: G. K. Gilbert, U.S. Geological Survey. Page 91, top: Montana Highway Commission; bottom: Charles A. Knell, Bureau of Reclamation. Page 92: Los Angeles Times. Page 96: Teledyne Geotronics. Page 97: Institut Geographique National. Page 98: P. Jay Fleisher, © 1972, Harper & Row. Page 99: Defense Intelligence Agency. Page 100: Corps of Engineers, U.S. Army New Orleans District. Page 103: French Government Tourist Office. Page 105: National Oceanic and Atmospheric Administration. Page 106: U.S. Army Corps of Engineers, July 2, 1969. Page 107: George A. Grant, National Park Service. Page 108: P. Jay Fleisher, © 1972, Harper & Row. **Chapter 5** Page 114: Courtesy of the American Museum of Natural History. Page 120, top: E. B. Hardin, U.S. Geological Survey; bottom: Courtesy of the American Museum of Natural History. Page 130: Kay Y. James. Page 131: Richard Frear, National Park Service. **Chapter 6** Page 142: Paul Almasy, Camera Press Limited. Page 145: Oregon State Highway. Page 146: Grant Institute of Geology, University of Edinburgh. Page 147: Geological Survey of Canada, Ottawa. Page 152: Swiss National Tourist Office. Page 153, top left: Cliff Bottomley, Australian Information Service; top right: M. Woodbridge Williams, National Park Service; bottom: U.S. Geological Survey.

Viewpoints

Chapter 1 Pages 20–23: Edited from "Of Man and Nature" in *Festschrift for Ronald Duncan*; Harold Lockyear, Editor. Harton Press. **Chapter 5** Pages 136–139: Adapted by permission of Alfred A. Knopf, Inc. from *The Closing Circle—Nature, Man and Technology*, by Barry Commoner. Copyright © 1971 by Barry Commoner. A substantial portion of this book originally appeared in *The New Yorker*.

Line Drawings

Chapter 1   Fig 1–5: After R. Garrels and F. Mackenzie, *Evolution of Sedimentary Rocks*, New York: Norton, 1971, p. 45. Fig 1–6: After A. Strahler, *Planet Earth*, New York: Harper & Row, 1972, p. 156. Fig 1–9: After M. K. Hubbert, *Energy Resources*, Washington, D.C.: National Academy of Sciences, 1962.   **Chapter 2**   Fig 2–1: After W. C. Ernst, *Earth Materials*, Englewood Cliffs, N.J.: Prentice-Hall, 1969, p. 10. Fig 2–6: After R. Garrels and F. Mackenzie, *Evolution of Sedimentary Rocks*, New York: Norton, 1971, p. 22.   **Chapter 3**   Fig 3–5: After S. Clark, *Structure of the Earth*, Englewood Cliffs, N.J.: Prentice-Hall, 1971, p. 7.   **Chapter 5**   Fig 5–2: After A. McAlester, *The History of Life*, Englewood Cliffs, N.J.: Prentice-Hall, 1968, p. 142.   **Chapter 6**   Fig 6–5: After *Investigating the Earth*, Earth Science Curriculum Project, Boston: Houghton Mifflin, 1972, p. 425. Fig 6–7: After R. Garrels and F. Mackenzie, *Evolution of Sedimentary Rocks*, New York: Norton, 1971, p. 122. Fig 6–9, Fig 6–10: After G. Wolman and J. Miller, "Magnitude and Frequency of Forces in Geomorphic Processes," *Journal of Geology*, vol. 68, 1960, pp. 56, 59.

Tables

Chapter 1   Table 1–4: After M. Calvin, *Chemical Evolution*, New York, Oxford University Press, 1969, pp. 107, 112.   **Chapter 3**   Table 3–1: Modified after A. Holmes, *Principles of Physical Geology*, New York: Ronald Press, 1965, p. 901.   **Chapter 6**   Table 6–2, Table 6–3, Table 6–4: After National Research Council, National Academy of Sciences, *The Earth and Human Affairs*, San Francisco: Canfield Press, 1972, pp. 41–42.

Index

Page numbers in **boldface** type indicate glossary entries.